SOLVING PARTIAL DIFFERENTIAL EQUATIONS ON PARALLEL COMPUTERS

SOLVING PARTIAL DIFFERENTIAL EQUATIONS ON PARALLEL COMPUTERS

JIANPING ZHU

Department of Mathematics and Statistics
NSF Engineering Research Center
Mississippi State University
Mississippi State, Mississippi 39762, USA

Published by

World Scientific Publishing Co. Pte. Ltd.
5 Toh Tuck Link, Singapore 596224
USA office: 27 Warren Street, Suite 401-402, Hackensack, NJ 07601
UK office: 57 Shelton Street, Covent Garden, London WC2H 9HE

British Library Cataloguing-in-Publication Data
A catalogue record for this book is available from the British Library.

SOLVING PARTIAL DIFFERENTIAL EQUATIONS ON PARALLEL COMPUTERS

ISBN-13 978-981-02-1578-1
ISBN-10 981-02-1578-9

To Yan and Lily.

Preface

It has been almost fifty years since the first computer was built on earth. During that period, scientific computing has made dramatic progresses in solving large scale complex problems in different application areas. It is now recognized that, in addition to theoretical analysis and experimentation, scientific computing has become the third means for exploring the nature. At the center of scientific computing is the research on numerical solution of partial differential equations describing complex physical problems. As the speed of computers increases, researchers are solving more and more complicated problems, and the demand in computer speed always outgrows the development in computer hardware. The attempt to solve large scale real-world problems has pushed the speed of single-processor computers to the limit. It is now clear that using multiple processors on a parallel computer is a cost-effective way, and probably the only way, to achieve the performance required for solving the so-called "Grand Challenge" problems.

Althongh the research in parallel computing has entered the second decade. We still have not seen the wide-spread use of parallel computers on a regular basis for solving large practical problems in application areas. On one hand, there are more and more active research projects being conducted by researchers in computer science and applied mathematics on parallel processing, including hardware, system software and numerical algorithms. New parallel computers have been emerging every two to three years, some of these machines became out of date even before they were used to solve any practical problems. On the other hand, there are growing interests in using parallel computers from researchers who work in practical application areas. They are not familiar with the programming methodologies in a multi-processor environment on parallel computers. The purpose of this book is to bridge the gap between these two groups, and to quickly bring researchers and graduate students working on numerical solutions of partial differential equations with various applications into the area of parallel processing.

The book is organized as follows: Chapter 1 gives a brief introduction to parallel computers and some basic concepts in parallel processing. Chapter 2 discusses several popular algorithms for solving PDEs on parallel computers, including basic relaxation algorithms, ADI algorithm, multigrid algorithm and conjugate gradient algorithm. In order to make the material more general, the discussions in this chapter are concentrated on the analyses of the algorithms without referring to a particular parallel architecture. The implementation issues are discussed in Chapter 3, and examples of solving the Poisson's equation and a multiphase flow problem are given in Chapter 4.

Although part of the materials in these two Chapters is machine-dependent and may become outdated as new architectures emerge, the general ideas discussed, like minimizing the ratio of communications to computations and overlapping communications with computations, are useful for a much wider class of parallel computers than was discussed in the book. I believe that the use of concrete examples can demonstrate better the procedures involved in implementing and fine-tuning programs on parallel computers, especially for the researchers who are new in this area. Chapter 5 discusses parallel time stepping algorithms which exploit parallelism in both the spatial and temporal dimensions. With these algorithms, solutions at different time steps are solved concurrently. Finally, Chapter 6 outlines the future development in high performance computing.

It is understandable that many important topics are just discussed very briefly, or even not covered, in a book of this size. I hope the book can be used as a starting point for people who are interested in using parallel computers to solve PDE related problems in their application areas. Many references are given in the book to direct interested readers to more specific topics and in-depth analyses.

This book project was initiated by an invitation from the World Scientific Publishing. I appreciate very much the opportunity to write this book, and the effort of the editors and staff at the World Scientific Publishing that made the publication of this book possible.

I am also deeply grateful to the National Science Foundation Enginnering Research Center and the Department of Mathematics and Statistics at Mississippi State University. Without the generous support from these units, it would not have been possible for me to finish this book project. Actually, some of the research work discussed in this book was funded and conducted at the Engineering Research Center on the campus of Mississippi State University. The encouragement and support from my friends and colleagues contributed significantly to this book project. They also provided me with many of the references collected in this book. In particular, Dr. Yuefan Deng of the State University of New York at Stony Brook read the manuscript and gave me many valuable suggestions and corrections.

The complete manuscript of this book, including all text, figures, and tables, was prepared using LaTeX and a few other graphics packages on Sparc workstations. It is by no means an easy task to typeset a book and to combine the text generated by LaTeX and the figures generated by other graphics packages. I am indebted to many of my students whose expertise in word-processing and graphics packages made it possible to produce the complete camera-ready manuscript using LaTeX. Special thanks go to Mr. Xiaobing Gu who typeset most of the text and figures in this book and spent countless nights working on the manuscript, and to Mr. Xinglai Zhuang

who typeset the first chapter and improved significantly the quality of many figures
in this book with his artistic talents. He also participated in the research projects
discussed in Chapter 3 and Chapter 4.

Finally, I want to thank my wife, daughter and parents for their support and
understanding which was essential to the completion of this book project. I wish I
had been writing this book without spending too much time on teaching and research
at the same time. Unfortunately, that was not the case. This book project was carried
out in a parallel processing style with my teaching, research, and other professional
activities simultaneously. For the past year, I spent almost all my spare time writing
this book. It is probably hard for other people to believe that, in that period, it was
considered as a great event by my wife and daughter if I took them to a McDonald's
fast food restaurant to have a BigMac and spent half an hour watching my daughter
playing in the restaurants playground. Such an event would keep my daughter excited
for days. I know this was a little too tough to them because I had taken away too
much time that I was supposed to spend with them. Now the project is over, I will
make it up as soon as possible and I am glad that they are all happy to see the book
coming out.

Starkville, Mississippi,   JIANPING ZHU
October, 1993.

Contents

Chapter

1

Introduction

1.1   Exploiting Parallelism

Parallelism exists everywhere in our society and the nature. For example there are many workers in an auto factory working together (in parallel) to produce cars; a university might have hundreds of faculty and staff members carrying out teaching and research activities simultaneously; and a student, while listening to a lecture, might be at the same time thinking about how to have a fun party for the weekend. All these are instances where there are more than one activity, either working or thinking, going on in parallel. In the case of production lines in a factory, the purpose of exploiting parallelism by using many workers is to increase the productivity or output. There are two important factors that will affect the output of these parallel processes. The first is that everyone involved must be doing some useful work simultaneously so that the whole task can be completed in a shorter time by using more workers. The second is that there must be proper and efficient coordinations among all workers. Without proper coordination, the whole factory will be a mess. On the other hand, if there is too much coordination or inefficient coordination, then one can not expect high efficiency in the production lines.

Parallel computing for solving large scale scientific and engineering problems is also characterized by the same factors discussed above. The rationale of exploiting parallelism in scientific computing is to increase the throughput of a computer by making it do more than one operation at the same time.

A traditional sequential computer has one central processing unit (CPU) that handles basic numerical operations sequentially, i.e. one operation at a time. For large scale scientific problems, the amount of numerical computations involved is so large that even the fastest sequential computers can not produce solutions in a reasonable amount of time. There are basically two ways to make a computer work faster: either to make a faster CPU so that an operation will take less time to finish, or to put more than one CPU in a computer so that multiple operations can be done simultaneously. The latter is called parallel processing because there are multiple processing units, usually called processors, working in parallel on a computer. Due to the difficulties in wire fabrication and the limit of the speed of light, it is getting more and more difficult to build faster single-processor computers. On the other hand, it is relatively easier to put a number of processors (not necessary the fastest) together so that the aggregate performance is significantly increased. The research in parallel computing, both in hardware and software, has made dramatic progresses in the last decade [44, 95]. However, to fully realize the computing power of parallel computers, we must make all processors on a parallel computer busy doing useful work, and coordinate properly the computations on different processors to ensure that the final results are correct.

1.2   Parallel Computers

A typical single-processor computer is shown in Figure 1.1, where both instructions

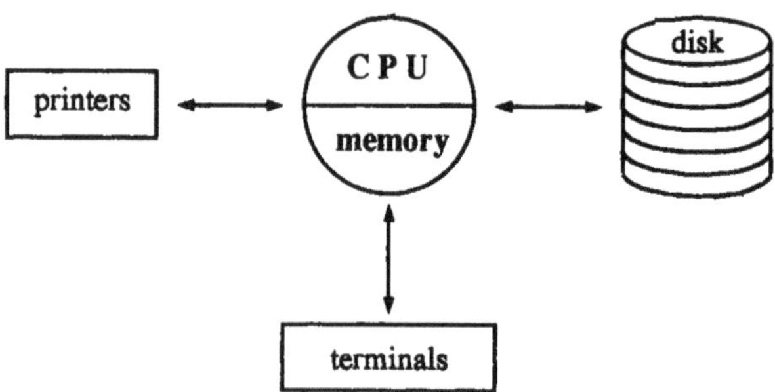

Figure 1.1: A single-processor computer

and data are stored in the main memory. The CPU fetches instructions and data from the memory, performs the operations specified by the instructions and puts the results of the computations back to the memory.

A parallel computer has multiple processors that execute the same or different instructions on different data sets. There are many different kinds of parallel computers that can be classified into different categories based on their memory access or execution control patterns.

1.2.1 Shared Memory and Distributed Memory.

Based on memory access patterns, parallel computers have been classified into two main categories: Shared memory parallel computers and distributed memory parallel computers.

1.2.1.1 Shared Memory Parallel Computers.

The configuration of a shared memory parallel computer is shown in Figure 1.2. All processors have access to the same main memory. Different instructions can be

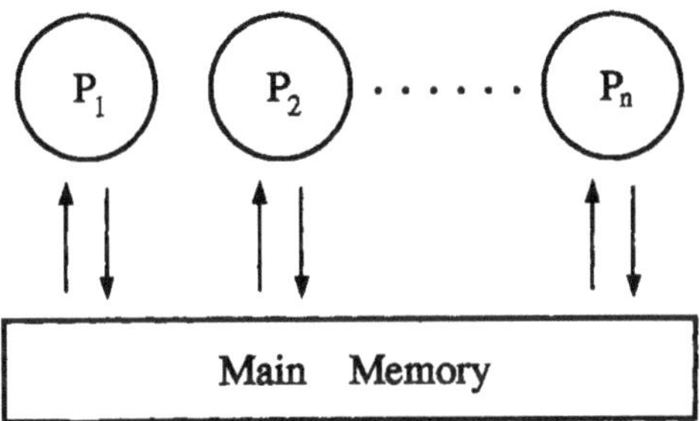

Figure 1.2: A shared memory parallel computer, P_i represents different processors.

executed by different processors in parallel. Most of today's working-horse supercomputers fall into this category. Typical representatives include Cray Y-MP, Cray C90, NEC SX/3, Hitachi S820, and IBM3090. These machines usually have from one to sixteen processors, several hundred megabytes or several gigabytes of memory, and gigaflops (10^9 floating point operations per second) level performance. Prices range from several to tens of millions of dollars.

1.2.1.2 Distributed Memory Parallel Computers.

The configuration of a distributed memory parallel computer is shown in Figure
1.3. Each processor has access to its own local memory and exchanges data with

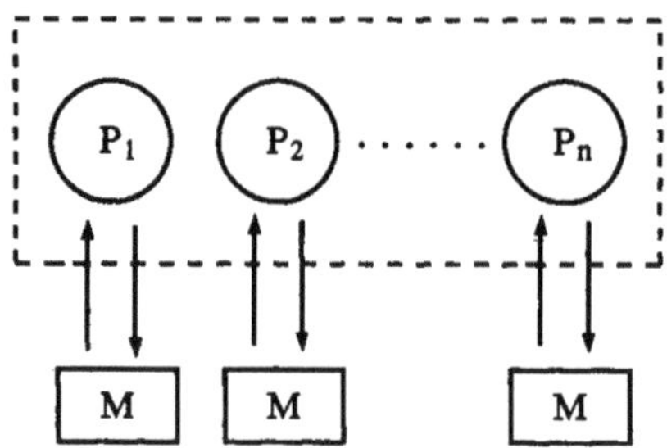

Figure 1.3: A distributed memory parallel computer, P_i represents a processor and M_i represents the local memory.

other processors by inter-processor communications. The frame encompassing all
processors in Figure 1.3 represents connections between different processors. Typical
representatives include Paragon, iPSC/860, CM5, nCUBE2 and Maspar. There are
many different interconnection topologies for distributed memory parallel computers.
Figure 1.4 shows a ring connection in which a processor is connected to its two nearest
neighbors. Interprocessor communications are realized by exchanging data between
processors. The speed of communication channels (the bandwidth) is defined as the
number of bytes of data transmitted per second between two nearest neighboring
processors. It is obvious that computational tasks requiring local data exchange
(say nearest neighbor exchange) can make better use of the communication channels
between all processors.

A mesh connection is shown in Figure 1.5, where each processor is connected to
four processors considered as neighbors on the north, south, east and west. On some
architectures there may not be the wrap-around connections between the first column
and the last column, the first row and the last row, respectively. If the length of the
communication path between two nearest processors is defined as 1, the maximum
length of the communication path between any two processors in a square mesh of P
processors is $O(\sqrt{P})$.

Another special connection topology is the hypercube connection in which the
number of processors P is usually a power of 2, say $P = 2^d$. Each processor is

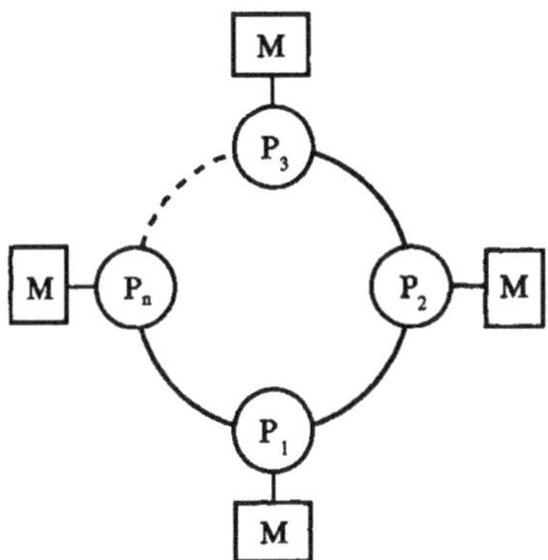

Figure 1.4: Ring connection, P_i represents a processor and M represents the local memory.

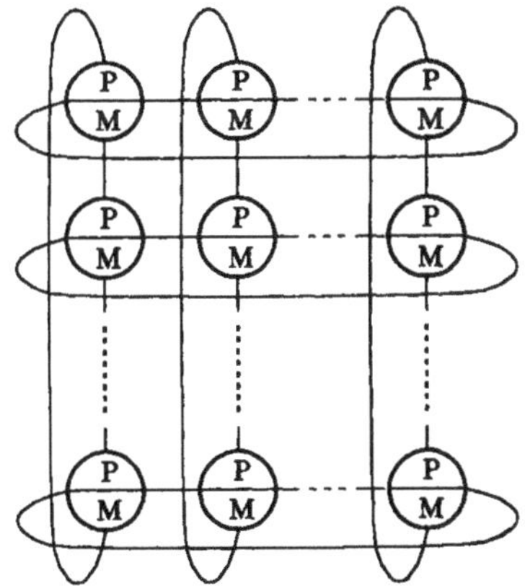

Figure 1.5: Mesh connection, P represents a processor and M represents the local memory.

connected to d processors and the resulting connection is called a d-dimensional hypercube. Figure 1.6 shows a 3-dimensional hypercube with eight processors.

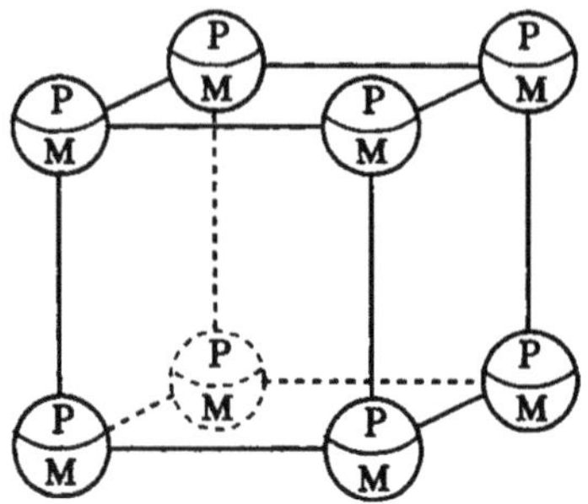

Figure 1.6: Hypercube connection, P represents a processor and M represents the local memory. Each processor on this three-dimensional hypercube is connected to three other processors.

The maximum length of the communication path between any two processors on a d-dimensional hypercube with $P = 2^d$ processors is d. More detailed discussions about hypercubes will be given later in Chapter 3.

The number of processors on a distributed memory parallel computer ranges from tens to tens of thousands. The processing speed of each processor is relatively slow (a few to tens of megaflops) compared with that of shared memory machines. Prices vary significantly based on the number of processors and the size of local memory on each processor, from several hundred thousand dollars to millions of dollars.

1.2.2   SIMD and MIMD

Parallel computers can also be divided into different categories based on execution control patterns, i.e. whether different processors can execute different instructions simultaneously.

1.2.2.1   SIMD Parallel Computers.

On a SIMD parallel computer (SIMD stands for Single Instruction Multiple Data), all processors execute the same instruction on different data sets. The Connection machine (like CM2) is a typical representative in this category. With tens of thousands of processors, a CM2 is very suitable for homogeneous computations.

For example, the addition of two matrices A and B can be expressed as

$$
\begin{bmatrix} c_{11} & \cdots & c_{1n} \\ \vdots & & \vdots \\ c_{n1} & \cdots & c_{nn} \end{bmatrix} = \begin{bmatrix} a_{11} & \cdots & a_{1n} \\ \vdots & & \vdots \\ a_{n1} & \cdots & a_{nn} \end{bmatrix} + \begin{bmatrix} b_{11} & \cdots & b_{1n} \\ \vdots & & \vdots \\ b_{n1} & \cdots & b_{nn} \end{bmatrix},
$$

where the basic operation is $c_{ij} = a_{ij} + b_{ij}$. If there are n^2 processors arranged as a $n \times n$ grid, then processor P_{ij} can be assigned to perform $c_{ij} = a_{ij} + b_{ij}$. Therefore all processors perform the same operation of adding two numbers simultaneously, the only difference is that different processors are adding two different numbers.

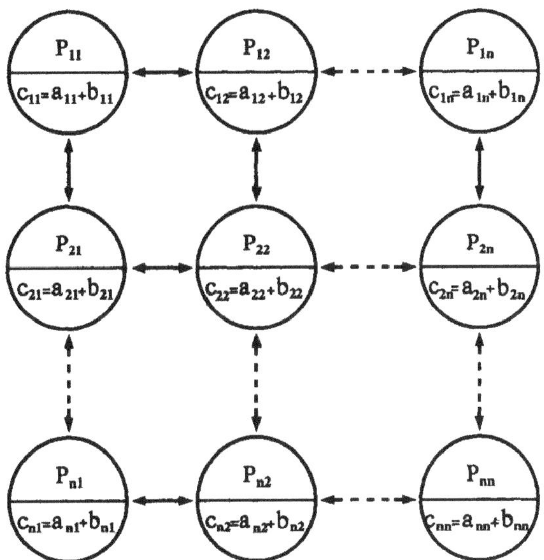

Figure 1.7: Addition of two matrices on a SIMD parallel computer, processor P_{ij} performs $c_{ij} = a_{ij} + b_{ij}$.

For large matrices, we can decompose the original matrices into submatrices and assign submatrices to processors. All processors will then perform the same operations simultaneously on different submatrices.

1.2.2.2 MIMD Parallel Computers.

the processors on a MIMD (MIMD stands for Multiple Instruction Multiple Data) parallel computer are able to execute different operations on different data sets simultaneously. For example, while one processor is performing an operation of adding two vectors, the other processors might be performing the inner product of another two vectors, or multiplication of a scalar with a vector at the same time, as shown in Figure 1.8.

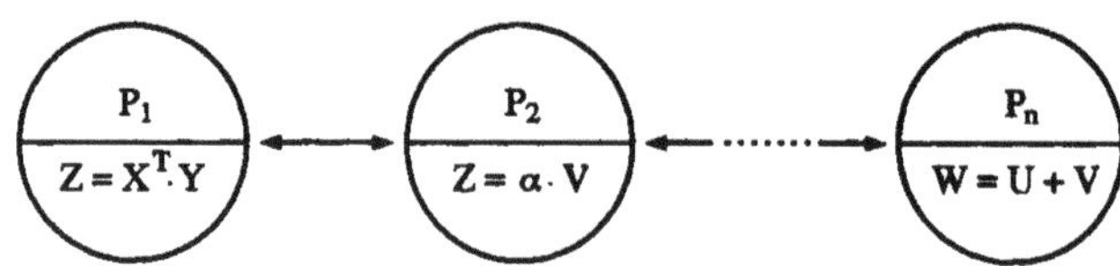

Figure 1.8: Different processors P_i on a MIMD parallel computer can perform different operations simultaneously.

Most parallel computers fall into this category, including shared memory parallel computers, like Cray and IBM3090, and distributed memory parallel computers like nCUBE and Intel hypercubes.

1.2.3 Shared Address Space and Distributed Physical Memory

For application programmers, it is easier to develop programs on shared memory parallel computers than on distributed memory parallel computers, since all processors have direct access to all the data located in the common memory on parallel computers with shared memory. However, it is difficult to scale this type of architectures to a large number of processors with proportionally increased memory because of the limit on bandwidth (the rate at which data can be transferred between the processors and the shared memory). This is the major reason why there are usually only a small number of processors on a shared memory parallel computer.

On the other hand, distributed memory parallel computers can be scaled to a large number of processors with proportionally increased aggregate memory. For example, on an Intel iPSC/860 hypercube, each processor can have up to 64 megabytes of memory. Therefore a 32-processor hypercube will have 2 gigabytes of memory and a 128-processor hypercube will have 8 gigabytes of memory. However, for application programmers, the algorithm implementation is more complicated since a processor has access only to its own local memory of up to 64 megabytes. If a data item located in another processor is needed, it must be transferred to the requesting processor by using special communication mechanisms. It is the programmers responsibility to find out when a data item on one processor is needed by another processor and to arrange the data exchange through communication channels.

To combine the advantages of both the shared memory and distributed memory parallel computers, researchers have developed another architecture on which users feel that there is a shared memory accessible by all processors, so they do not have to handle explicitly data distributions and communications. But the physical memory is actually distributed over different processors and each processor only has direct access to its local memory. Therefore, the architecture is still scalable to a large number of processors. Figs. 1.9, 1.10, and 1.11 show how a big array A is stored on different

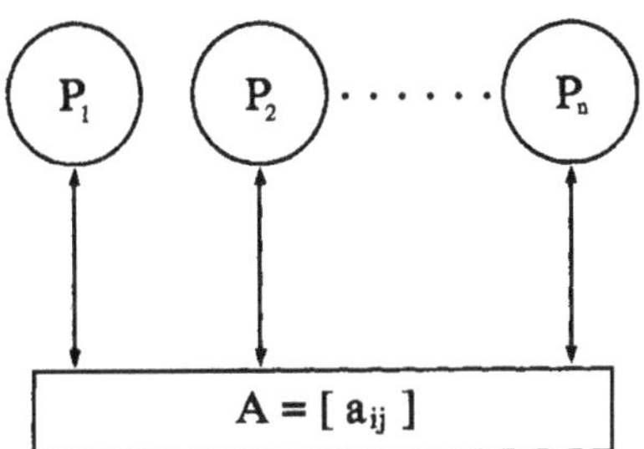

Figure 1.9: A big array $A = (a_{ij})$, $i, j = 1, 2, \cdots, n$ is stored in a physically shared memory which is accessible directly by all processors.

architectures.

Figure 1.10 shows how a big array $A = [A_1, A_2, \cdots, A_n]$, where A_i's are submatrices, is stored on a distributed memory computer. Each processor has direct access to a submatrix of the original matrix. In the event processor P_i needs a data item in submatrix A_j which is stored in the local memory of P_j, interprocessor communication between P_i and P_j must be used in the user program to move the data item.

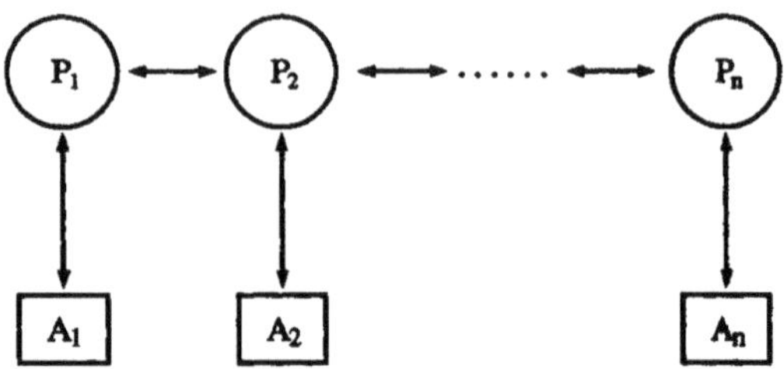

Figure 1.10: A big array A is decomposed into submatrices and stored in distributed local memory of different processors.

Figure 1.11 shows the environment of shared address space and distributed physi-

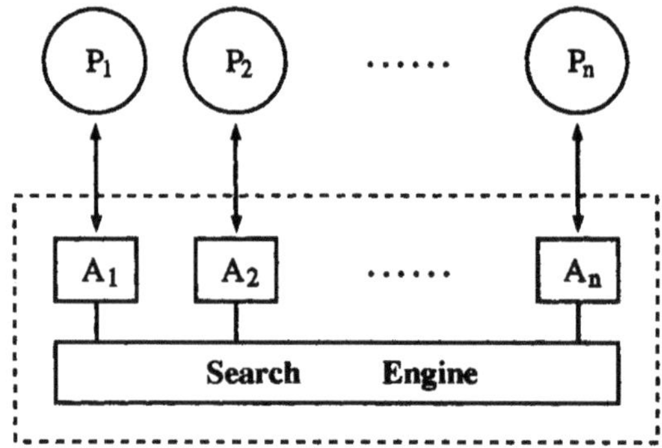

Figure 1.11: A big array A is stored in a parallel computer with shared address space and distributed physical memory.

cal memory and how the big array A is stored in it. The picture is very similar to the situation with distributed memory parallel computers, except there is a search engine which interconnects the local memory of all processors and provides routing and directory services. An algorithm performing operations on array A can be implemented on a processor P_i in the manner as if the whole array $A = (a_{ij}), i, j = 1, 2, \cdots, m,$ were stored in the local memory of processor P_i. When a data item is referenced which is not in the local memory of processor P_i, the search engine will locate the data and bring the data into the local memory of processor P_i. As a result, the physically distributed memory behaves logically as a single shared address space, in which all processors can access a particular data item, although the physical memory is distributed over different processors. This architecture is also called the shared virtual memory parallel computer.

The KSR-1 parallel computer produced by the Kendall Square Research is a parallel computer with shared address space and distributed physical memory. Each processor on a KSR-1 has 32 megabytes of local memory, a search engine connects local memory of different processors. High scalability is achieved by using a hierarchy of search groups as shown in Figure 1.12. The bandwidth requirement to the next

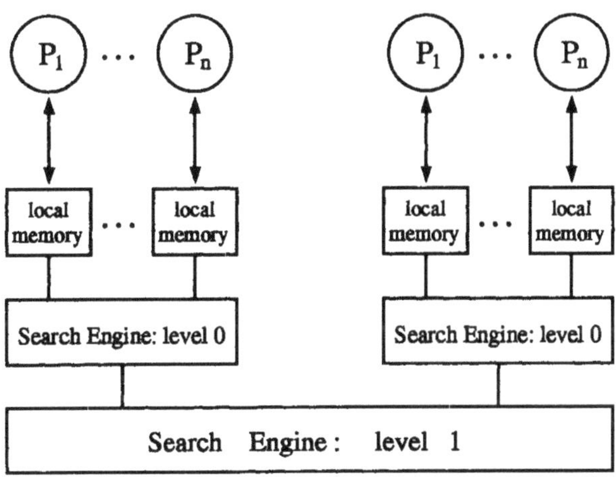

Figure 1.12: Hierarchy of search groups on KSR-1.

higher level can be reduced by satisfying most references from within the level where the reference request is initiated. Because of its scalability and user-friendliness, many researchers believe that the architecture with shared address space and distributed physical memory is very promising in the future [45].

1.3   Performance Evaluations

1.3.1   Speedup

For a given problem and a parallel computer with P processors, the speedup is defined as

$$S = \frac{T_S}{T_P} \tag{1.1}$$

where T_S is the wallclock time required to solve the given problem by the best known sequential algorithm on one processor and T_P is the wallclock time required to solve the problem by a parallel algorithm on P processors. Another commonly used definition for speedup is

$$S = \frac{T_1}{T_P} \tag{1.2}$$

where T_1 and T_P are the wallclock times used by the same parallel algorithm to solve the given problem on one and P processors, respectively. It is obvious that to get consistent speedup measures, one has to have an exclusive access to the parallel computer being used. Otherwise, depending on the system load, the times T_1 and T_p will be different for different runs if the computer is shared by many users.

Intuitively, the speedup S of a parallel algorithm running on a parallel computer measures the reduction of the computing time originally used by the sequential algorithm on one processor for solving the same problem. A speedup of $S = 5$ means the solution can be obtained with multiple processors in 1/5 of the original time required by a single processor computer.

It should be noted hat the speedups calculated using (1.1) and (1.2) are usually different because T_s and T_1 are usually different. To take advantage of multiple processors on a parallel computer, we may have to restructure the sequential algorithm developed for uni-processor computers and schedule as many computations in parallel as possible. Because of these structure changes and re-arrangement of computations, the best sequential algorithm and the parallel algorithm may require a different amount of computations for solving the same problem. Therefore, we usually have $T_1 \geq T_S$ in general because the parallel algorithm running on one processor is not necessarily the best sequential algorithm. The numerical complexity of a parallel algorithm is usually higher than that of the sequential algorithm for the same problem. For example, the sequential algorithm for solving equation

$$
\begin{bmatrix}
a_1 & & & & \\
c_1 & a_2 & & & \\
& c_2 & \ddots & & \\
& & \ddots & \ddots & \\
& & & c_{N-1} & a_N
\end{bmatrix}
\begin{bmatrix}
x_1 \\
x_2 \\
\vdots \\
x_N
\end{bmatrix}
=
\begin{bmatrix}
b_1 \\
b_2 \\
\vdots \\
b_N
\end{bmatrix}
$$

can be written as

$$
x_1 = \frac{b_1}{a_1}
$$

$$
x_i = \frac{(b_i - c_{i-1} x_{i-1})}{a_i} \qquad i = 2, \cdots, N
$$

which requires approximately $3N$ operations. One way to parallelize this sequential solution process is to use the cyclic reduction algorithm. For $n = 8$, the reduction process is as follows:

        stage 1   **stage 2**   **stage 3**

$$
a_1 x_1 = b_1
$$
$$
\left. c_1 x_1 + a_2 x_2 = b_2 \right\} \qquad d_2 x_2 = e_2
$$
$$
c_2 x_2 + a_3 x_3 = b_3 \qquad\qquad\qquad \left.\right\} \qquad g_4 x_4 = h_4
$$
$$
\left. c_3 x_3 + a_4 x_4 = b_4 \right\} \quad f_2 x_2 + d_4 x_4 = e_4
$$
$$
\qquad\qquad\qquad\qquad\qquad\qquad\qquad \left.\right\} \quad k\, x_8 = b
$$
$$
\left. c_4 x_4 + a_5 x_5 = b_5 \right\} \quad f_4 x_4 + d_6 x_6 = e_6
$$
$$
c_5 x_5 + a_6 x_6 = b_6 \qquad\qquad\qquad \left.\right\} \quad k_4 x_4 + g_8 x_8 = h_8
$$
$$
\left. c_6 x_6 + a_7 x_7 = b_7 \right\} \quad f_6 x_6 + d_8 x_8 = e_8
$$
$$
c_7 x_7 + a_8 x_8 = b_8
$$

There are three stages involved in the process. In each stage, the number of equations is halved by merging a pair of equations into a single equation. The mergings of different pairs of equations within a stage are independent to each other and can be executed in parallel.

After x_8 has been solved from the final equation, the value can be injected backward to solve the rest of the unknowns. Again, the backward solution processes for different unknowns within a stage are independent to each other and can be done in parallel. If there are totally N equations, then $\log_2 N$ stages are needed to reduce them into a single equation and then solve all unknowns backwards respectively. A straightforward implementation of this algorithm requires approximately $7N$ operations, which can be reduced to $5N$ if some quantities are calculated once and stored for later use. Either way, the numerical complexity of the cyclic reduction algorithm is significantly higher than that of the sequential algorithm which is $3N$.

In the ideal case, the best sequential algorithm is also the parallel algorithm. For example, in the matrix addition example of section 1.2.2.1, the sequential algorithm takes n^2 additions to add two $n \times n$ matrices. Since the addition processes for different entries are independent, the same algorithm can also be executed on parallel computers without increasing the numerical complexity of the algorithm.

In general, however, a parallel algorithm for solving a given problem has higher numerical complexity than that of the best sequential algorithm for the same problem. As long as the reduction of computing time by using multiple processors is more than what is needed to offset the increase of computing time due to the higher numerical complexity of the parallel algorithm, it is worthwhile using parallel computers so that solutions can be calculated faster.

1.3.2   Efficiency

The efficiency of a parallel algorithm running on a parallel computer with P processors is defined as

$$e = \frac{S}{P}$$

where S is the speedup. The efficiency e measures the effective utilization of all processors by a parallel algorithm, or the portion of the time a processor is doing useful work for solving the given problem during the algorithm execution. The higher the efficiency, the better a parallel computer is utilized by a parallel algorithm. In an ideal case, if the algorithm is perfectly parallelizable, one would expect a P−fold speedup if P processors are used, or in other words, the problem can be solved in $1/P$ of the original time required by one processor. In this case, the efficiency will be

$$e = \frac{S}{P} = \frac{P}{P} = 1$$

which means all processors are busy all the time in the execution doing useful work for solving the given problem. In general, however, the speedup S is less than P due to the parallel overhead, higher numerical complexity, synchronizations, and non-parallelizable part of the problem. Therefore, we usually have $e < 1$.

1.3.3 Amdahl's law

The purpose of using multiple processors is to reduce the turnaround time for solving a given problem. As discussed in the previous sections, a P-fold speedup means that the problem can be solved in $1/P$ of the original time used by one processor. It is natural that people expect to achieve higher speedup by using more processors. However, since all computations for solving a given problem are usually not parallelizable and a parallel algorithm usually has higher numerical complexity than that of the sequential algorithm, one can not expect a $P-$ fold speedup by using P processors. Actually, for a given problem with a fixed size, there is a bound on the speedup expressed as the Amdahl's law [6]. Suppose N operations are needed to solve a given problem and τ is the time required by a processor to perform one operation, then the sequential execution time on one processor will be

$$T_1 = N\tau.$$

If a fraction α $(0 \leq \alpha \leq 1)$ of the N operations are parallelizable, meaning αN operations can be performed in parallel, then the execution time on P processors for solving the same problem will be

$$T_P = (1 - \alpha)N\tau + \frac{\alpha N\tau}{P}.$$

By definition (1.2), the speedup is then

$$S = \frac{T_1}{T_P} = \frac{N\tau}{(1 - \alpha)N\tau + \frac{\alpha N\tau}{P}} = \frac{1}{(1 - \alpha) + \frac{\alpha}{P}} = \frac{P}{(1 - \alpha)P + \alpha}. \tag{1.3}$$

Unless $\alpha = 1$, we have $S < P$, which means using P processors will not result in a $P-$ fold speedup. If (1.3) is written as

$$S = \frac{1}{(1 - \alpha) + \frac{\alpha}{P}},$$

we can see that S is bounded by $\frac{1}{1-\alpha}$ which is the highest achievable speedup when $P \rightarrow \infty$.

Figure 1.13 shows the curves of speedup S as a function of the number of processors P for different values of α. When $\alpha = 1$, which means all computations are parallelizable, we see the ideal linear speedup $S = P$. If $\alpha = 0.99$, which means only 1 percent of all operations is non-parallelizable, the speedup curve soon deviates from the ideal linear speedup. The highest speedup possible in this case is $\frac{1}{1-\alpha} = \frac{1}{1-0.99} = \frac{1}{0.01} = 100$. The situation with $\alpha = 0.75$ is even more striking: No

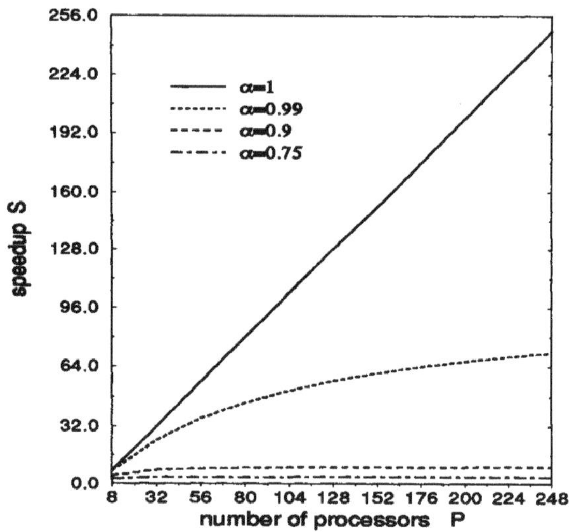

Figure 1.13: Speedup curves vs. the number of processors P and different values of α.

matter how many processors are used in the computation. The speedup S is always less than 4! This demonstrates the significant impact of the non-parallelizable part on the performance of parallel computers. The central part of the research in parallel algorithm development is to maximize α so that higher speedup can be achieved.

1.3.4 Gustafson's Model

The results of the previous section might appear to be very disappointing since the speedup is bounded by the sequential part of an algorithm. However, the results shown in Figure 1.13 are based on the assumption that the problem size is fixed while the number of processors increases. If we analyze the speedup S in a different setting, the situation will be quite different. Considering the fact that parallel computers are developed for solving large scale problems, not for testing problems which can be solved on both one and P processors, we should increase the problem size as the number of processors P increases. The fraction α of the parallelizable computation is not a constant in general. In many cases, α increases as the problem size increases.

There are two ways to generalize the definition of speedup given by (1.3) so that it can be used to describe the performance of a parallel computer for problems of increasing sizes. One way is to introduce a new model to describe the speedup. Gustafson [73] proposed an alternative model based on fixing the computation time, rather than the problem size. Assume that a given problem can be solved in one unit of time on a parallel machine with P processors and that the times in sequential and parallel computations are $1 - f$ and f, respectively, then the time for a uniprocessor computer to solve the same problem would be $1 - f + fP$. The speedup $S_{P,f}$ is thus given by

$$S_{P,f} = \frac{1 - f + fP}{1} = fP + 1 - f \tag{1.4}$$

We can see that $S_{P,f}$ is a linear function of P and scales up in a different way than that in Eq. (1.3) as the number of processors P increases.

The other way is to consider the fact that α, the portion of parallelizable computation, usually increases as the size of problems increases. Therefore, even with the speedup definition given by (1.3), the actual speedup will not be limited in the way as shown in Figure 1.13. To see that this is true, consider solving the Laplace equation

$$\frac{\partial^2 u}{\partial x^2} + \frac{\partial^2 u}{\partial y^2} = 0 \qquad\qquad (x,y) \in D$$

$$u = f(x,y) \qquad\qquad (x,y) \in \partial D$$

defined on a square domain D as shown in Figure 1.14

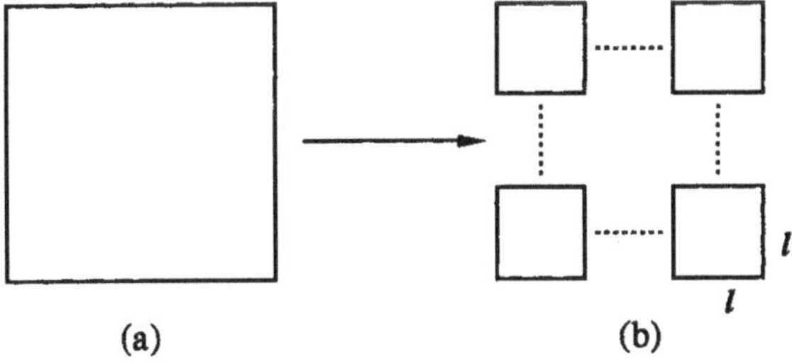

Figure 1.14: (a) the original domain, (b) the decomposed domain for parallel computing

If the Jacobi iterative algorithm and the domain decomposition method are used, which will be discussed in details in Chapters 2 and 3, the domain shown in figure 1.14(a) should be decomposed into patches as shown in figure 1.14(b). Each processor

takes one patch and calculates the solution in that patch. The computational parts
are perfectly parallel. However, after each iteration, solutions on the boundary of
a patch must be exchanged with the neighboring patches held by other processors,
this part does not contribute directly to the calculation of the solution. Assume that
each subdomain has dimensions $l \times l$ with l^2 grid points. The amount of data in
a patch to be exchanged is proportional to the length of the circumference of the
patch which is $4l$, while the amount of computation in a patch is proportional to the
area of the patch which is l^2. Therefore, the ratio of the parallelizable part over the
non-parallelizable part is

$$r = \frac{l^2}{4l} = \frac{l}{4} \ .$$

As the size l of patches increases, so does the ratio r, which means that the portion
of parallelizable work will increase as l increases. As a result, the speedup can not be
fully described by a single curve shown in Figure 1.13. Instead, as the problem size
increases, the speedup should be given by curves with higher and higher α values.

Chapter

2

Parallel Algorithms for Solving Partial Differential Equations

Many physical and engineering phenomena are described by partial differential equations (PDEs),which can be written in a general form as

$$L(u) \quad = \quad f(\underline{x}, t) \qquad \underline{x} \in \Omega \qquad t > 0 \qquad (2.1)$$

$$B(u) \quad = \quad g(\underline{x}, t) \qquad \underline{x} \in \partial\Omega \qquad t > 0 \qquad (2.2)$$

$$U(\underline{x}, 0) \quad = \quad h(\underline{x}, 0) \qquad \underline{x} \in \Omega \qquad (2.3)$$

where L and B are differential operators, $f(\underline{x}, t)$, $g(\underline{x}, t)$ and $h(\underline{x}, 0)$ are given functions, Ω is the given spatial domain, and $\partial\Omega$ is the boundary of Ω. Eqs. (2.2) and (2.3) represent boundary and initial conditions. Since Eqs. (2.1)–(2.3) can not be solved analytically for most practical problems, numerical methods are very important for the computation of approximate solutions. In this book, we will assume that Ω is either a rectangular two-dimensional or a cubical three-dimensional domain. For more complicated domains, special methods have to be used to discretize the relevant PDEs or spatial domains. For example, in computational fluid dynamics, complex geometries which are referred to as the physical domains are transformed into rectangular and cubical domains by numerical grid generation procedures. These transformed regular domains are usually referred to as the computational domains. The use of the regular computational domains greatly simplifies the discretization of PDEs and the implementation of numerical algorithms. The trade off is that the original PDE usually becomes more complicated after the transformation. The simplest example of such a transformation is probably the polar coordinate transformation. As shown in Fig. 2.1, the Laplace equation with the Dirichlet boundary condition on a circular domain in the Cartesian coordinates x-y will have the form of

$$\frac{\partial^2 u}{\partial x^2} + \frac{\partial^2 u}{\partial y^2} = 0 \qquad r_1 < \sqrt{x^2 + y^2} < r_2 \tag{2.4}$$

$$u(x,y) = g_1(x,y) \qquad \sqrt{x^2 + y^2} = r_1,$$

$$u(x,y) = g_2(x,y) \qquad \sqrt{x^2 + y^2} = r_2,$$

where r_1 and r_2 are the radia of the two boundary circles, $g_1(x,y)$ and $g_2(x,y)$ are given functions. We can see that since the boundary conditions are given in terms of x and y, it is not convenient to use in the computations. Fig. 2.1 shows that the polar coordinate transformation

$$x = r \cos \alpha$$
$$y = r \sin \alpha$$

maps the circular domain in the x-y coordinate system onto a rectangular domain

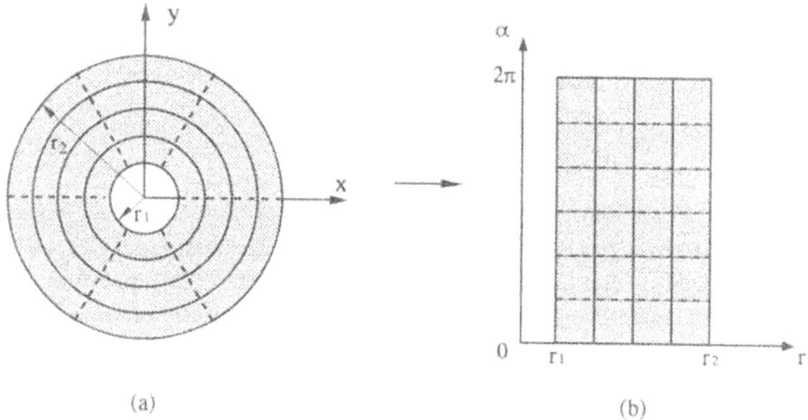

Figure 2.1: Polar coordinate transformation maps a circular domain in (a) onto a rectangular domain in (b).

in the r-α coordinate system. In the transformed coordinate system, we have a new PDE that is defined on a rectangular computational domain:

$$\frac{\partial^2 u}{\partial r^2} + \frac{1}{r}\frac{\partial u}{\partial \alpha} + \frac{1}{r^2}\frac{\partial^2 u}{\partial \alpha^2} = 0, \qquad r_1 < r < r_2, \qquad 0 \le \alpha \le 2\pi,$$

$$u(r, \alpha) = g_1(\alpha), \qquad\qquad r = r_1,$$

$$u(r, \alpha) = g_2(\alpha), \qquad\qquad r = r_2,$$

$$u(r, 0) = u(r, 2\pi), \qquad\qquad r_1 < r < r_2.$$

We can see that with the rectangular computational domain shown in Fig. 2.1(b), the boundary conditions are in a very simple form. Although the original PDE becomes more complicated in the transformed domain, the benefit of having a rectangular domain for algorithm implementation is often more than what is needed to offset the increased complexity of the PDEs in practical situations. Therefore, the numerical grid generation method is widely used in applications to map the complex physical domains onto regular computational domains [172].

In general, partial differential equations are classified into three categories: elliptic, parabolic, and hyperbolic. For detailed theoretical analysis on analytic solution methods, the existence, uniqueness, and stability of solutions, readers are referred to [41, 59]. For most application problems, it is impossible to obtain a solution using analytical methods. Therefore, numerical solution methods have played an important role in solving scientific and engineering problems governed by partial differential equations. Although the earliest application of the finite difference method to the numerical solutions of partial differential equations dated back to the beginning of this century [151, 153], the method was not widely used in practical applications in early days. It was the advent of digital computers in the 40s and 50s that made numerical methods powerful tools for solving partial differential equations. Since then, there has been continued development in different numerical algorithms for solving PDEs, including finite difference, finite element, finite volume, and spectral methods, and the list keeps growing. Detailed discussions on these methods can be found in [7, 117, 164]. The basic procedure involved in all these numerical solution methods is to obtain a system of algebraic equations by discretizing either the spatial and temporal domains, or the solution space (a functional space). The numerical solutions are then obtained by solving these algebraic equations.

In this chapter, we will concentrate on discussing parallel algorithms for solving the linear algebraic equation systems resulting from the discretization of PDEs. Two simple PDEs, the Poisson's Equation and the heat equation, are used first as examples to demonstrate the discretization procedures using the finite difference scheme. The rest of this chapter is devoted to the discussions of parallel relaxation method, ADI method, multigrid method, and conjugate gradient method.

2.1   Discretization by Finite Difference Method

1. **Poisson's equation**: Many important physical phenomena, in particular the static field problems like the electromagnetic field and the incompressible potential flow field, are described by elliptic PDEs. A typical representative is the Poisson's equation

$$\frac{\partial^2 u}{\partial x^2} + \frac{\partial^2 u}{\partial y^2} = f(x,y) \qquad (x,y) \in \Omega \tag{2.5}$$

with either the Dirichlet condition

$$u(x,y) = g(x,y) \qquad (x,y) \in \partial\Omega$$

or the Neumann condition

$$\frac{\partial u}{\partial n} = h(x,y) \qquad (x,y) \in \partial\Omega$$

specified on the boundary, where g and h are given functions and $\frac{\partial u}{\partial n}$ is the normal derivative. In more general cases, both types of conditions can appear in the same problem with the Dirichlet condition specified on one part of the boundary and the Neumann condition specified on the other part.

To obtain the numerical solution of Eq. (2.5), the rectangular domain Ω should first be discretized by a set of grid points as shown in Fig. 2.2. There are different ways to number all grid points in the discretized domain. Fig. 2.2(a) and 2.2(b) show two ways of numbering the grid points in the domain using either one or two indices respectively. The original PDE is also discretized so that we have an algebraic equation for each grid point. The grid indexing scheme shown in Fig. 2.2(b) is very convenient for expressing different discretization schemes for a given PDE, while the indexing scheme in Fig. 2.2(a), which is often called the Lexicographic ordering, is more convenient for assembling the matrix equations over the whole domain.

We now use the finite difference method to discretize Eq. (2.5). For an interior grid point P_{ij}, the second order partial derivatives $\frac{\partial^2 u}{\partial x^2}$ and $\frac{\partial^2 u}{\partial y^2}$ can be discretized by the central finite difference formulas

$$\left.\frac{\partial^2 u}{\partial x^2}\right|_{i,j} = \frac{u_{i-1,j} - 2u_{i,j} - u_{i+1,j}}{\Delta x^2}$$

and

$$\left.\frac{\partial^2 u}{\partial y^2}\right|_{i,j} = \frac{u_{i,j-1} - 2u_{i,j} - u_{i,j+1}}{\Delta y^2}$$

respectively. At each grid point, the discretized form of Eq. (2.5) is

$$\frac{u_{i-1,j} - 2u_{i,j} + u_{i+1,j}}{\Delta x^2} + \frac{u_{i,j-1} - 2u_{i,j} + u_{i,j+1}}{\Delta y^2} = f(x_i, y_j).$$

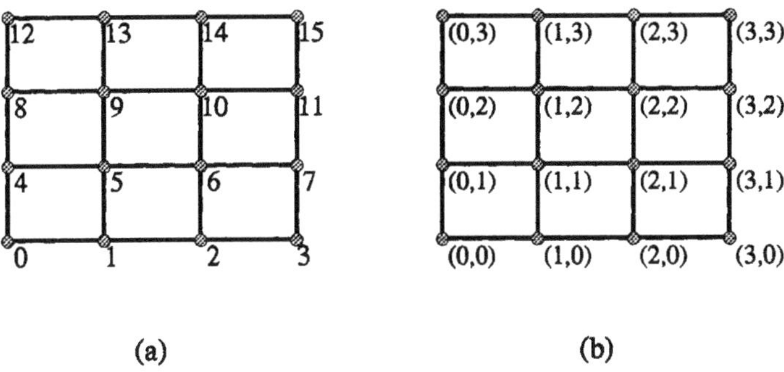

Figure 2.2: A rectangular domain Ω discretized by the finite difference method. (a) Grid points are numbered using lexicographic ordering. (b) Grid points are numbered using two indices.

If the grid spacing is uniform with $\Delta x = \Delta y = h$, then we have

$$u_{i,j-1} + u_{i-1,j} - 4u_{i,j} + u_{i+1,j} + u_{i,j+1} = h^2 f(x_i, y_j) \tag{2.6}$$

$$i = 1, 2, \cdots NI, \quad j = 1, 2, \cdots NJ$$

where there are $NI + 2$ and $NJ + 2$ points in the x and y directions numbered from 0 to $NI + 1$ and 0 to $NJ + 1$ respectively. For the grid points that connect directly to one or more boundary points, there will be one or more known terms on the left hand side of Eq. (2.6) since $u(x, y)$ is given on the boundary. In this case, the known terms should be moved to the right hand side of the equation. As mentioned before, the single subscript indexing is more convenient for assembling equations over the whole domain. Therefore, we denote $u_{i,j}$ as u_l, where $l = (j - 1) \times NI + i$, when assembling the linear algebraic equations. The assembled equation is

$$Au = b \tag{2.7}$$

where

$$A = \begin{bmatrix} A_1 & I & & & \\ I & A_2 & \ddots & & \\ & \ddots & \ddots & I & \\ & & I & A_{NJ} \end{bmatrix},$$

and

$$
A_i = \left[\begin{array}{ccccc}
-4 & 1 & & & \\
1 & -4 & \ddots & & \\
& \ddots & \ddots & 1 & \\
& & 1 & -4 &
\end{array} \underbrace{\hphantom{xxxxxx}}_{NI} \right] .
$$

Eq. (2.7) is a penta-diagonal system with $NI \times NJ$ equations in the system. The major work of solving the Poisson's Equation given in Eq. (2.5) numerically is to solve the algebraic equation system (2.7).

For a three-dimensional problem

$$
\frac{\partial^2 u}{\partial x^2} + \frac{\partial^2 u}{\partial y^2} + \frac{\partial^2 u}{\partial z^2} = f(x,y,z), \qquad (x,y,z) \in \Omega,
$$

with the Dirichlet boundary condition

$$
u(x,y,z) = g(x,y,z), \qquad (x,y,z) \in \partial\Omega,
$$

where Ω is a regular cubic domain, we can similarly discretize the domain by a set of $(NI+2) \times (NJ+2) \times (NK+2)$ grid points. Assume all grid points are uniformly distributed in all three dimensions with a grid spacing h, we have at each grid point an equation similar to Eq. (2.6):

$$
u_{i,j,k-1} + u_{i,j-1,k} + u_{i-1,j,k} - 6u_{i,j,k} + u_{i+1,j,k} + u_{i,j+1,k} + u_{i,j,k+1} = h^2 f(x_i, y_j, z_k)
$$

$$
i = 1, 2, \cdots NI, \quad j = 1, 2, \cdots NJ, \quad k = 1, 2, \cdots NK.
$$

Assembling equations at all grid points together with one-subscript indexing $l = (k-1) \times NI \times NJ + (j-1) \times NI + i$, we will have an algebraic system similar to Eq. (2.7) with

$$
A = \left[\begin{array}{ccccc}
A_1 & I & & & \\
I & A_2 & \ddots & & \\
& \ddots & \ddots & I & \\
& & I & A_{NK} &
\end{array} \right], \tag{2.8}
$$

$$
A_k = \left[\begin{array}{ccccc}
A_1^k & I & & & \\
I & A_2^k & \ddots & & \\
& \ddots & \ddots & I & \\
& & I & A_{NJ}^k &
\end{array} \right],
$$

$$
A_j^k \;=\; \begin{bmatrix}
-6 & 1 & & \\
1 & -6 & \ddots & \\
& \ddots & \ddots & 1 \\
& & 1 & -6
\end{bmatrix} \cdot
$$

$$\underbrace{\hphantom{-6 \quad 1 \quad \quad 1 \quad -6}}_{NI}$$

The coefficient matrix A is now 7-diagonal and there are totally $NI \times NJ \times NK$ equations in the system.

For more general equations like

$$
\frac{\partial}{\partial x}\left(b_1 \frac{\partial u}{\partial x}\right) + \frac{\partial}{\partial y}\left(b_2 \frac{\partial u}{\partial y}\right) + \frac{\partial}{\partial z}\left(b_3 \frac{\partial u}{\partial z}\right) = f(x,y,z)
$$

where b_i's are functions of (x,y,z), we can still obtain a discretized algebraic equation system with similar structure if the central finite difference scheme is used. The major difference is that the elements in the coefficient matrix are no longer 1 or -6, instead we may have

$$
A \;=\; \begin{bmatrix}
A_1 & D_1 & & \\
D_1 & A_2 & \ddots & \\
& \ddots & \ddots & D_{NK-1} \\
& & D_{NK-1} & A_{NK}
\end{bmatrix}
$$

$$
D_k \;=\; \begin{bmatrix}
d_1^k & & & \\
& d_2^k & & \\
& & \ddots & \\
& & & d_{NI \times NJ}^k
\end{bmatrix}
$$

$$
A_k \;=\; \begin{bmatrix}
A_1^k & B_1^k & & \\
B_1^k & A_2^k & \ddots & \\
& \ddots & \ddots & B_{NJ-1}^k \\
& & B_{NJ-1}^k & A_{NJ}^k
\end{bmatrix}
$$

where D_k and B_j^k are diagonal matrices of order $NI \times NJ$ and NI respectively, and A_j^k is a tri-diagonal matrix of order NI.

2. **Heat Equation**: This equation is a typical representative of the time dependent partial differential equations:

$$
\frac{\partial u}{\partial t} = \frac{\partial^2 u}{\partial x^2} + \frac{\partial^2 u}{\partial y^2} + \frac{\partial^2 u}{\partial z^2} + f(x,y,z,t), \tag{2.9}
$$

$$(x, y, z) \in \Omega, \qquad\qquad t > 0,$$

$$u(x, y, z, t) = g(x, y, z, t), \qquad (x, y, z) \in \partial\Omega, \qquad t > 0,$$

$$u(x, y, z, 0) = h(x, y, z), \qquad (x, y, z) \in \Omega.$$

The right hand side of Eq. (2.9) can be discretized in a similar way as for Eq. (2.5), so that we have

$$\frac{\partial u}{\partial t} = Au + f \tag{2.10}$$

where $u = \{u_1, u_2, \cdots\cdots u_N\}^T$, $f = \{f_1, f_2, \cdots\cdots f_N\}^T$ and $N = NI \times NJ \times NK$. The coefficient matrix A in Eq. (2.10) has the same structure as that shown in Eq. (2.8). If a one step implicit finite difference scheme is used to discretize the time derivative in Eq. (2.10), we will have

$$\frac{u^n - u^{n-1}}{\Delta t} = Au^n + f^n, \qquad n = 1, 2, \cdots NT, \tag{2.11}$$

where u^n represents the solution at the discretized time level t_n, and $\Delta t = t_n - t_{n-1}$. Eq. (2.11) can be re-written as

$$(I - \Delta t A)u^n = u^{n-1} + \Delta t f^n \qquad n = 1, 2, \cdots NT, \tag{2.12}$$

or as

$$Bu^n = b^n, \qquad n = 1, 2, \cdots NT, \tag{2.13}$$

with

$$B = I - \Delta t A, \qquad b^n = u^{n-1} + \Delta t f^n.$$

The coefficient matrix B in Eq. (2.13) still has the same structure as that of A in (2.8). Therefore, the solution algorithms used for solving Eq. (2.7) can all be used for solving Eq. (2.13). The major difference is that Eq. (2.13) represents many algebraic equation systems at different time levels t_n, $n = 1, 2, \cdots NT$. The solution process usually starts from $n = 1$ where the right hand side is known since u^0 is given as the initial condition. Once u^1 is solved, it can be put into the right hand side of Eq. (2.12) for solving u^2. The process repeats until the desired solutions at all time steps are solved. Since the process starts from the initial condition u^0 and marches step by step through the time levels where solutions are desired, it is called the time marching or the time stepping algorithm. Eqs. (2.7) and (2.13) show that the central part for solving linear elliptic or parabolic PDEs is to solve a linear algebraic equation system. In the case of parabolic PDEs, there is one algebraic equation system to be solved at each time step.

The solution algorithms for solving Eqs. (2.7) and (2.13) can be divided into two categories: The direct method and the iterative method. Since the iterative method is suitable for large, sparse matrix equations resulting from discretizations of PDEs (linear or nonlinear) on complex spatial domains, we will mainly discuss developments and implementations of iterative algorithms on parallel computers for solving Eqs. (2.7) and (2.13). Readers are referred to [29, 45, 143, 144] for discussions of fast direct algorithms for solving the same equations on parallel computers.

2.2 Parallel Relaxation Algorithms

All relaxation type algorithms can be written as

$$u_k = Tu_{k-1} + c, \qquad k = 1, 2, \cdots,$$

where u_0 is the initial guess used to start the iterations. Based on how the iterative matrix T is constructed from the coefficient matrix A in the equation $Au = b$, we will have the classic Jacobi, Gauss-Seidel and SOR relaxation algorithms.

2.2.1 Jacobi Algorithm

The Jacobi relaxation algorithm begins by decomposing the coefficient matrix A into $A = D - L - U$, where D, L and U are the diagonal, lower triangular and upper triangular parts of A, respectively. The iterations can be expressed as

$$Du_k = (L + U)u_{k-1} + b, \qquad k = 1, 2, \cdots,$$

or in the component form

$$u_i^{(k)} = (b_i - \sum_{j=1, j \neq i}^{n} a_{ij} u_j^{(k-1)})/a_{ii}, \qquad (2.14)$$

$$i = 1, 2, \cdots n, \qquad k = 1, 2, \cdots.$$

For a particular case like Eq. (2.7), we have

$$\begin{pmatrix} u_1 \\ u_2 \\ \vdots \\ u_N \end{pmatrix}^k = \frac{-1}{4} \begin{bmatrix} 0 & -1 & & -1 & & \\ -1 & 0 & \ddots & & \ddots & \\ & \ddots & \ddots & \ddots & & -1 \\ -1 & & \ddots & \ddots & \ddots & \\ & \ddots & & \ddots & \ddots & -1 \\ & & -1 & & -1 & 0 \end{bmatrix} \begin{bmatrix} u_1 \\ u_2 \\ \vdots \\ u_N \end{bmatrix}^{k-1} - \frac{1}{4} \begin{bmatrix} b_1 \\ b_2 \\ \vdots \\ b_N \end{bmatrix} \qquad (2.15)$$

$$k = 1, 2, \cdots.$$

where u_0 is the initial guess of the true solution u. To see more clearly how the computation in each iteration proceeds, we write Eq. (2.15) in component form with two-subscript indexing for a two-dimensional rectangular domain:

$$u_{i,j}^{(k)} = \frac{u_{i,j-1}^{(k-1)} + u_{i-1,j}^{(k-1)} + u_{i+1,j}^{(k-1)} + u_{i,j+1}^{(k-1)} - h^2 f(x_i, y_j)}{4} \qquad (2.16)$$

$$i = 1, 2, \cdots NI, \quad j = 1, 2, \cdots NJ, \quad k = 1, 2, \cdots.$$

Eq. (2.16) shows that the solution value at a point (i, j) is calculated, in each iteration, by using the solution values at the four neighboring points from the previous iteration. Therefore, the new solution value at a particular grid point depends on the old solution values at the four neighboring points. If different discretization schemes other than the central finite difference is used to discretize Eq. (2.5), and a different relaxation scheme is used to solve Eq. (2.7), then the data dependence will be different, as will be discussed shortly. The analysis of the data dependence is very important for implementing these algorithms on parallel computers.

2.2.2   Gauss-Seidel and SOR Algorithms

2.2.2.1   Gauss-Seidel Algorithm

Eq. (2.14) shows that in the Jacobi algorithm all $u_i^{(k)}$'s are calculated using the values $u_j^{(k-1)}$, $j = 1, 2, \cdots n$, from the previous iteration. Actually, the values $u_j^{(k)}$, $j = 1, 2, \cdots i-1$, have been updated when we start calculating $u_i^{(k)}$. These updated values are used to calculate $u_i^{(k)}$ in the Gauss-Seidel algorithm:

$$u_i^{(k)} = (b_i - \sum_{j=1}^{i-1} a_{ij} u_j^{(k)} - \sum_{j=i+1}^{n} a_{ij} u_j^{(k-1)})/a_{ii}, \qquad (2.17)$$

$$i = 1, 2, \cdots n, \qquad k = 1, 2, \cdots.$$

The matrix representation of this algorithm is

$$(D - L)u_k = U u_{k-1} + b \qquad k = 1, 2, \cdots,$$

and the counterpart of Eq. (2.16) is

$$u_{i,j}^{(k)} = \frac{u_{i,j-1}^{(k)} + u_{i-1,j}^{(k)} + u_{i+1,j}^{(k-1)} + u_{i,j+1}^{(k-1)} - h^2 f(x_i, y_j)}{4} \qquad (2.18)$$

$$i = 1, 2, \cdots NI, \quad j = 1, 2, \cdots NJ, \quad k = 1, 2, \cdots.$$

It is natural to expect that the Gauss-Seidel algorithm converges faster than the Jacobi algorithm because the latest available data is used in the computation. While the conclusion is not true for linear algebraic equations with general coefficient matrices, the Gauss-Seidel algorithm does converge faster (twice as fast asymptotically) than the Jacobi algorithm for algebraic equations derived from discretization of the Poisson's equation and the heat equation. For detailed analysis of relaxation algorithms, readers are referred to [183, 194].

2.2.2.2 SOR Algorithm (Successive Over-Relaxation Algorithm)

Examing Eqs. (2.14) and (2.17), we can see that the calculation of a new value $u_i^{(k)}$ at a point i depends on values $u_j^{(k)}$ or $u_j^{(k-1)}$ at the points other than point i. The value $u_i^{(k-1)}$ is not utilized in the calculation.

The central idea of the SOR algorithm is to compute a better approximation to the true solution by forming a linear combination of the current updated solution $u_i^{(k)}$ and the solution $u_i^{(k-1)}$ from the previous iteration:

$$\hat{u}_i = (b_i - \sum_{j=1}^{i-1} a_{ij} u_j^{(k)} - \sum_{j=i+1}^{n} a_{ij} u_j^{(k-1)})/a_{ii},$$

$$u_i^{(k)} = (1-\omega)u_{i-1}^{(k-1)} + \omega \hat{u}_i. \tag{2.19}$$

The matrix representation of this algorithm can be written as

$$(D - \omega L)u_k = [(1-\omega)D + \omega U]u_{k-1} + \omega b, \tag{2.20}$$

and for the particular case of Eq. (2.7), we have

$$u_{i,j}^{(k)} = (1-\omega)u_{i,j}^{(k-1)} + \omega \frac{u_{i,j-1}^{(k)} + u_{i-1,j}^{(k)} + u_{i+1,j}^{(k-1)} + u_{i,j+1}^{(k-1)} - h^2 f(x_i, y_i)}{4}, \tag{2.21}$$

$$i = 1, 2, \cdots NI, \quad j = 1, 2, \cdots NJ, \quad k = 1, 2, \cdots.$$

It is clear from Eqs. (2.19) and (2.20) that the SOR algorithm reduces to the Gauss-Seidel algorithm when $\omega = 1$. The SOR algorithm is convergent when $0 < \omega < 2$ for systems with positive definite symmetric matrices. The selection of the parameter ω is difficult and important for the convergence rate. Only in the very special case, for example positive definite matrices resulting from the discretized Poisson's equation, can we have an analytic expression for ω [183, 194].

In contrast to Eq. (2.14), Eqs. (2.17) and (2.19) demonstrate a different type of data dependence: The computations of $u_i^{(k)}$ depend on the latest updated values

$u_j^{(k)}$, $j = 1, 2, \cdots i - 1$, and the values from the previous iteration $u_j^{(k-1)}$, $j = i, \cdots n$.. In the case of Poisson's equation on a two-dimensional rectangular domain with lexicographic ordering, Eqs. (2.18) and (2.21) show that the computations of $u_{ij}^{(k)}$ depend on the latest updated values at the western and southern neighboring points and the values from the previous iteration at the eastern and northern points. In the SOR algorithm, $u_{ij}^{(k)}$ also depends on $u_{ij}^{(k-1)}$.

Closely related to the SOR algorithm is the SSOR (symmetric SOR) algorithm in which a complete iteration consists of two SOR steps. The first step is just the usual SOR iteration:

$$\hat{u}_i^{(k)} = (1 - \omega)u_i^{(k-1)} + \omega(b_i - \sum_{j=1}^{i-1} a_{ij}\hat{u}_j^{(k)} - \sum_{j=i+1}^{n} a_{ij}u_j^{(k-1)})/a_{ii}, \qquad (2.22)$$

$$i = 1, 2, \cdots n.$$

The second step is an SOR iteration applied to the values obtained from Eq. (2.22) in a reversed order:

$$u_i^{(k)} = (1 - \omega)\hat{u}_i^{(k)} + \omega(b_i - \sum_{j=1}^{i-1} a_{ij}\hat{u}_j^{(k)} - \sum_{j=i+1}^{n} a_{ij}u_j^{(k)})/a_{ii},$$

$$i = n, \ n - 1, \ \cdots, \ 1.$$

For some problems, the SSOR method converges faster than the SOR algorithm by a factor of two or greater.

2.2.3   Parallel Relaxation Algorithms

We now discuss parallel relaxation algorithms for solving PDEs on distributed memory parallel computers. As discussed in Chapter 1, we have to keep all processors on a parallel computer busy doing useful work in order to complete the solution process in a shorter time.

2.2.3.1   Parallel Jacobi Algorithm

As analyzed in Section 2.2.1, computations for $u_{i,j}^{(k)}$ with the Jacobi algorithm depend only on values from the previous iteration. Fig. 2.3 shows the relation between the values of iteration k and those of iteration $k - 1$.

Since all values at iteration $k - 1$ should have been calculated before the kth iteration starts, all values at the kth level can be calculated independently and simultaneously. Therefore, the Jacobi algorithm is inherently parallel.

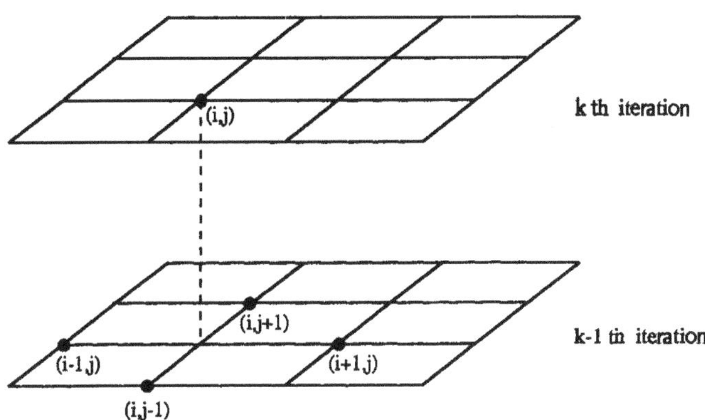

Figure 2.3: Computations of $u_{i,j}^{(k)}$ depend on values from the previous iteration at the points marked by •.

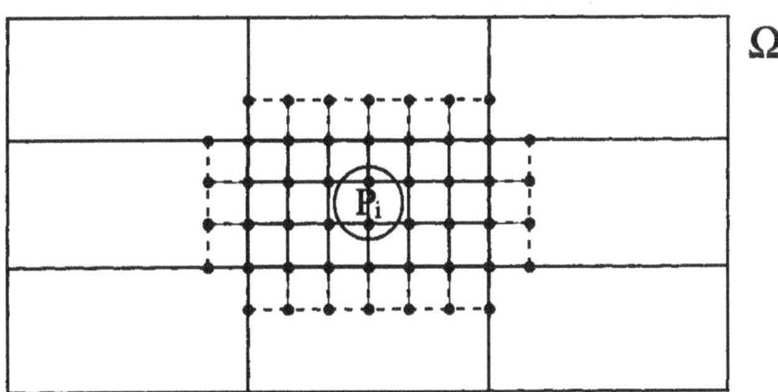

Figure 2.4: Processor P_i computes values at all points in a patch. Dashed lines indicate that the values at the points in other patches are needed to update all points on the boundary of the patch.

In practical situations, the number of grid points at which solutions are being calculated is usually much larger than the number of processors available on a parallel computer. Each processor is then assigned a subset of points to carry out the Jacobi iterations.

Fig. 2.4 shows that processor P_i is assigned all points in the patch bounded by the thick lines in the middle of the domain Ω. Since the relaxations of points on the boundary of the patch require values at the points on the dashed lines which are normally assigned to other processors, processor P_i must get values at those points, often called "ghost points", before it starts a new round of relaxation. Similarly, the values at the boundary of the patch must also be made available to other processors holding the neighboring patches at the end of each iterations.

The algorithm on one processor P_i can then be expressed as:

Algorithm 2.1
Parallel Jacobi Relaxations

initialize values at the "ghost points"
set error=10^{10}
while (error$> \varepsilon$)

1. Update all points using Eq (2.16).

2. Send values at the boundary of the patch to the neighboring processors.

3. Get updated values of the "ghost points" from the corresponding neighboring processors.

4. Compute the error.

end

where ε is the tolerance parameter. Steps 2 and 3 in the above algorithm are accomplished by interprocessor communications which are architecture dependent and will be discussed in more details in later chapters. The computation of the error in step 4 also requires information exchanges among processors since the iterations will terminate only when the maximum value of all errors at different processors is less than the given tolerance.

2.2.3.2   Parallel Gauss-Seidel Algorithm and Red-Black Ordering

The parallelization of the Gauss-Seidel algorithm is not as straightforward as that of the Jacobi algorithm as discussed in the previous section. As can be seen from Eq. (2.18), the computation of $u_{i,j}^{(k)}$ requires not only the values of $u_{i+1,j}^{(k-1)}$ and $u_{i,j+1}^{(k-1)}$ from the previous iteration, but also the values of $u_{i-1,j}^{(k)}$ and $u_{i,j-1}^{(k)}$ from the current iteration, as demonstrated by Fig. 2.5. Because of this dependence, the computation

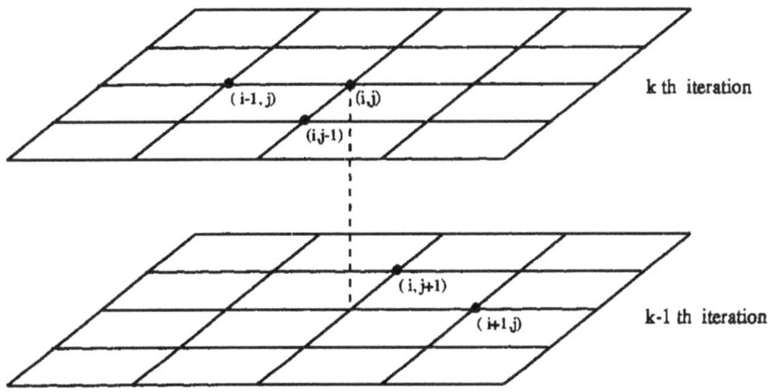

Figure 2.5: Computations of $u_{i,j}^{(k)}$ depend on the values from both the $k-1th$ iteration and the kth iteration at the points marked by •

for $u_{i,j}^{(k)}$ can not start until $u_{i-1,j}^{(k)}$ and $u_{i,j-1}^{(k)}$ have been calculated.

One way to parallelize the computations of the kth iteration is to use the wavefront ordering as shown in Fig. 2.6. The computation of the new values of iteration k follows the orders given in Fig. 2.6(b). We can see that all values for the points located on the same diagonal line can be calculated in parallel. The computation process proceeds like a wavefront. The correspondence between the usual 2-index ordering and the wavefront ordering is that the points with the same $i+j$ values are on the same diagonal. For example, all points on the highlighted front of Fig. 2.6(b) have $i+j=6$. The down side of this method is that the degree to which computations can be parallelized is not uniform. It varys in different stages of the computation. At the first diagonal in Fig. 2.6(b), only one value at point 1 can be calculated, the two values on the second diagonal can be calculated in parallel for points 2 and 3

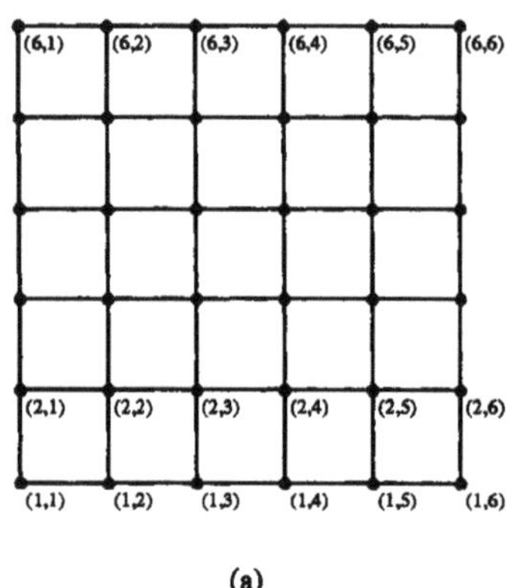
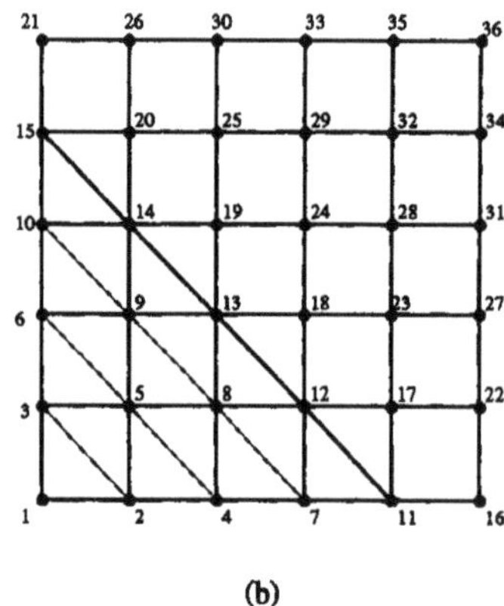

(a)   (b)

Figure 2.6: (a) 2-index ordering,   (b) the wavefront ordering

respectively. The degree of parallelism increases as the front moves forward until it passes the longest diagonal. After that, the degree of parallelism decreases as the front moves forward. Because of this nonuniform parallelism, it is hard to balance the computations on different processors and to achieve high efficiency.

Another more commonly used method to parallelize the Gauss-Seidel iteration is to use red-black ordering, or the even-odd ordering as shown in Fig 2.7. We can see

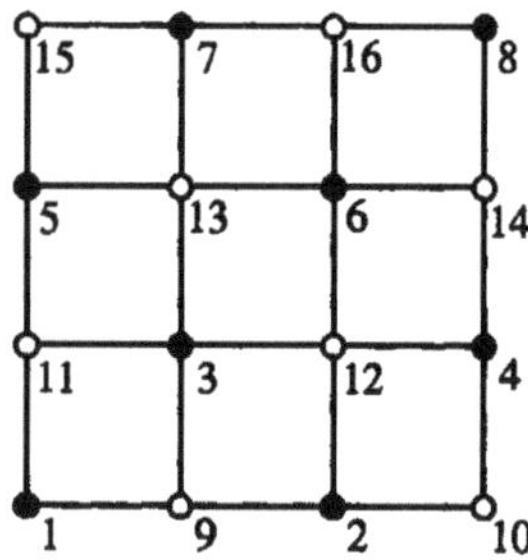

Figure 2.7: Red-black ordering, • black points, o red points

from Fig 2.7 that a point is marked either red or black. The coloring is done in such a way that no point is connected directly with another point of the same color. The ordering starts from the points of one color, say the black points, and then continues

to the points of another color. If we consider Eqs. (2.5) and (2.6) with this ordering
on the grid in Fig 2.7, then the coefficient matrix A in

$$Au = b \qquad (2.23)$$

will have the form

$$A = \begin{bmatrix} D_b & E \\ E^T & D_r \end{bmatrix}$$

where D_b and D_r are 8×8 diagonal matrices with -4 on the main diagonal and E
has the form of

$$E = \begin{bmatrix}
1 & & 1 & & & & & \\
1 & 1 & & 1 & & & & \\
1 & & 1 & 1 & 1 & & & \\
& 1 & & 1 & & 1 & & \\
& & 1 & & 1 & & 1 & \\
& & & 1 & 1 & 1 & & 1 \\
& & & & 1 & & 1 & 1 \\
& & & & & 1 & & 1
\end{bmatrix}.$$

Applying the Gauss-Seidel iteration to Eq. (2.23), we have

$$\begin{bmatrix} D_b & 0 \\ E^T & D_r \end{bmatrix} \begin{bmatrix} u_b \\ u_r \end{bmatrix}^k = \begin{bmatrix} 0 & E \\ 0 & 0 \end{bmatrix} \begin{bmatrix} u_b \\ u_r \end{bmatrix}^{k-1} + \begin{bmatrix} b_b \\ b_r \end{bmatrix} \qquad (2.24)$$

where $u_b = \{u_1 \cdots u_8\}^T$ and $u_r = \{u_9 \cdots u_{16}\}^T$ are the black and red unknowns,
respectively.

Eq. (2.24) can be separated as

$$D_b u_b^{(k)} = E u_r^{(k-1)} + b_b, \qquad (2.25)$$

and

$$E^T u_b^{(k)} + D_r u_r^{(k)} = b_r. \qquad (2.26)$$

Since D_b is a diagonal matrix with -4 on the main diagonal, Eq. (2.25) can be written
as

$$u_b^{(k)} = -\frac{1}{4}[E u_r^{(k-1)} + b_b] \qquad (2.27)$$

which means all components in $u_b^{(k)}$ can be calculated simultaneously. After $u_b^{(k)}$ has
been calculated, the red unknowns $u_r^{(k)}$ can then be calculated simultaneously by
substituting $u_b^{(k)}$ into Eq. (2.26), i.e.

$$u_r^{(k)} = -\frac{1}{4}[b_r - E^T u_b^{(k)}]. \qquad (2.28)$$

For general cases, if we have a $n \times m$ two-dimensional grid, the grid points should be ordered alternately as shown in Fig. 2.7. Assume that the number of grid points $n \times m$ is even, then we have $\frac{n \times m}{2}$ black and red unknowns respectively. All black unknowns can be solved in parallel first, then the red unknowns can be solved in parallel. Therefore, there are two steps in a complete iteration and each processor should take certain number of both red and black unknowns. In the first step, all processors solve the black unknowns by using Eq. (2.27), followed by the second step in which all processors solve the red unknowns by using Eq. (2.28).

Note that if a black or red point is on the boundary of a subdomain held by a particular processor, then the updated value will have to be exchanged with processors holding adjacent subdomains after each step in one iteration.

As an example, assume that there are four processors on a parallel computer and the domain is shown in Fig. 2.7, then each processor P_i, $i = 0,\ 1,\ 2,\ 3$, is assigned two black and red points respectively as shown in Fig. 2.8. At the beginning , all

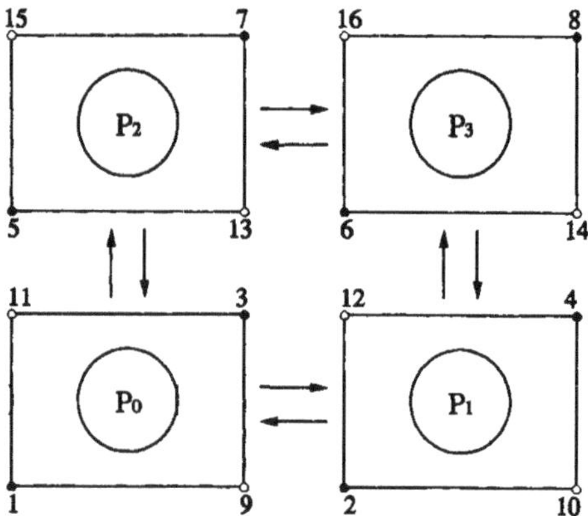

Figure 2.8: updated values at red and black points are exchanged after each step in one iteration. • black points, o red points.

unknowns are initialized using the initial guess. Therefore, all black unknowns can be updated simultaneously. For example, the value at point 2 can be updated by processor P_1 using the initialized values at points 9, 10, 12 and one value from the given boundary condition. Note that the initialized value at point 9 should also be available at processor P_1 since the value is predetermined before the start of the iteration. After all processors finish updating all black unknowns, the updated values must be sent to appropriate processors for updating the red unknowns. For example, the value at point 2 on processor P_1 must be sent to processor P_0 since the updating of red unknowns at point 9 depends on it. Similarly, processor P_1 must receive updated black unknowns at point 3 and point 6 from processors P_0 and P_3 respectively in order to update the red unknown at point 12. When all processors finish updating the red-unknowns, which completes one iteration, the updated values of red unknowns on the boundary of a subdomain should also be sent to the appropriate processors for updating the black unknowns in the next iteration. For example, processor P_1 should send the updated value at point 12 to processors P_0 and P_3 respectively for updating the black unknowns at point 3 and 6 respectively. It should also receive updated values at points 9 and 14 from processors P_0 and P_3 in order to update the black unknowns at points 2 and 4 in the next iteration.

The complete algorithm can be given as:

Algorithm 2.2
Parallel Gauss-Seidel Relaxations

initialize values at the "ghost points"
set error$=10^{10}$
while (error$> \varepsilon$)

1. update all black unknowns.

2. send those updated values on the boundary of the subdomain to appropriate processors.

3. receive updated black unknowns from appropriate processors to update the black ghost points.

4. update all red unknowns.

5. send the updated values on the boundary of the subdomain to appropriate processors.

6. receive updated red unknowns from appropriate processors to update the red ghost points.

7. calculate the error.

end

The SOR algorithm can be parallelized in exactly the same way except that Eq.

(2.27) and Eq. (2.28) should be written as

$$u_b^{(k)} = (1 - \omega)u_b^{(k-1)} - \frac{\omega}{4}[Eu_r^{(k-1)} + b_b], \tag{2.29}$$

and

$$u_r^{(k)} = (1 - \omega)u_r^{(k-1)} - \frac{\omega}{4}[b_r - E^T u_b^{(k)}], \tag{2.30}$$

respectively.

From Fig. 2.7, an intuitive explanation can be made about the Gauss-Seidel algorithm with red-black ordering. As shown in Fig. 2.7, the updating of all black points uses the values at the red points from the previous iterations, thus this step is similar to a Jacobi iteration applied to all black points. The updated values at the black points are then used in the second step to update all values at the red points which is similar to a Gauss-Seidel iteration applied to all red points. Therefore, a complete Gauss-Seidel iteration with red-black ordering actually consists of one Jacobi step and one Gauss-Seidel step. Because of the use of the Jacobi type iteration in the first step, the convergence rate of the colored Gauss-Seidel iteration is different from that of the conventional Gauss-Seidel iteration. More detailed analysis of the effect of coloring on convergence rate can be found in [2, 3, 115].

So far, all discussions about the parallel relaxation algorithms are based on two-dimensional domains for simplicity. Since the Jacobi algorithm is inherently parallel, it can be applied to equations on three-dimensional spaces without any change. For the Gauss-Seidel and SOR algorithms, the red-black ordering can still be used to parallelize the relaxation process. As shown in Fig. 2.9, a three-dimensional domain can be considered as a stack of two-dimensional domains. In each layer of a two-dimensional domain, the grid points are colored as red and black alternately. For different layers, the colors are just toggled so that no point is directly connected to a point of the same color in all three directions.

In practical situations, each processor will be assigned many red and black grid points. Instead of coloring a single grid point, we can actually color a block or a line of grid points so that unknowns in different blocks or lines of the same color can be updated simultaneously. These coloring schemes correspond to the block and line relaxation methods.

Note that so far we have assumed that a 5-point (7 point in three-dimensional domain) stencil has been used to discretize the given PDE as shown in Fig. 2.10. If higher order derivatives exist in the given PDE or if a higher order finite difference scheme is used, then the stencils shown in Fig. 2.11 might be used to discretize the original equation. In these cases, two colors are not enough to separate unknowns to make the diagonal blocks in Eq. (2.23) diagonal matrices. If the diagonal blocks

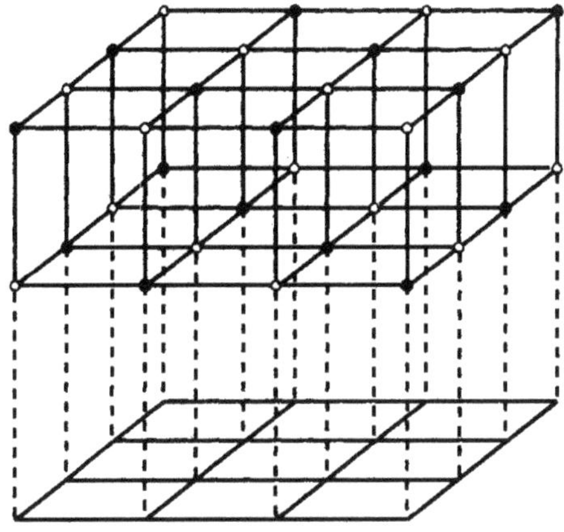

Figure 2.9: red-black ordering in a three-dimensional domain, the colors are toggled at two different layers in the third dimension. • black points, ○ red points.

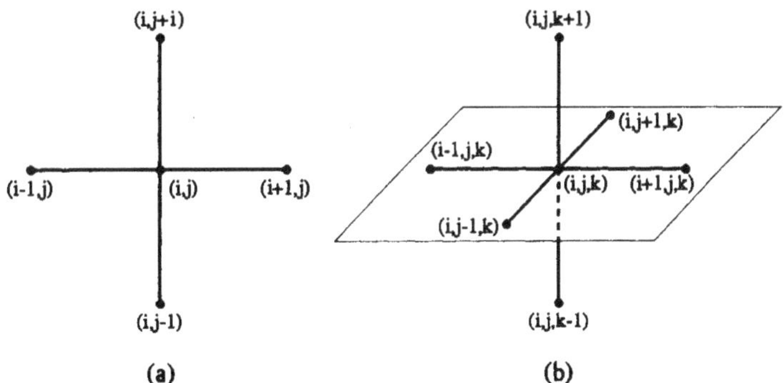

Figure 2.10: (*a*) A five-point stencil for the central finite difference in two-dimensional domains. (*b*) A seven-point stencil for three-dimensional domains.

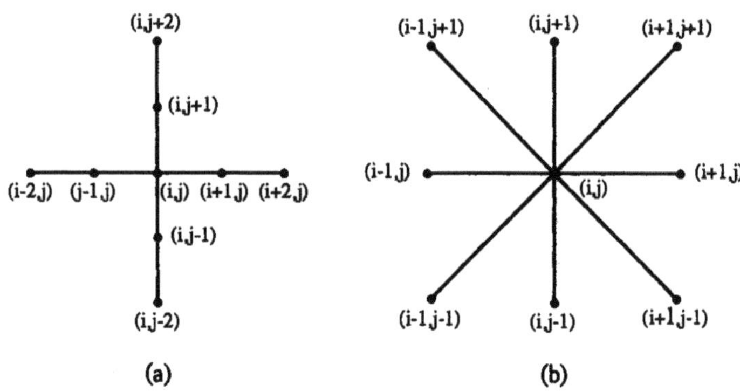

Figure 2.11: (a) A 9-point stencil for discretizing $\frac{\partial^2 u}{\partial x^2} + \frac{\partial^2 u}{\partial y^2}$ by the 4th order central finite difference. (b) A 9-point stencil for discretizing $\frac{\partial^4 u}{\partial x^2 \partial y^2}$ by the second order central finite difference.

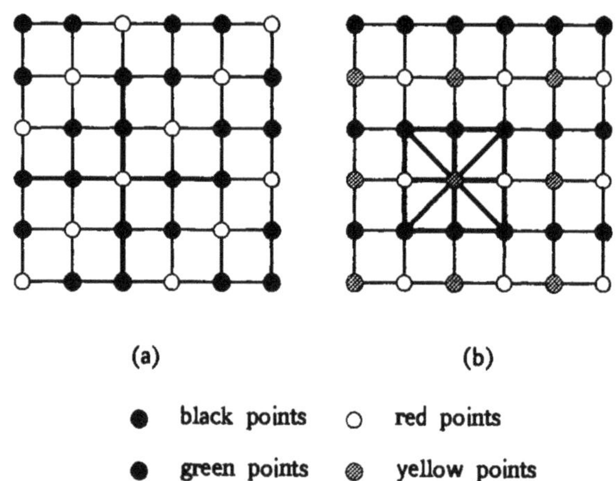

Figure 2.12: (a) A three-color ordering for the stencil shown in Fig. 2.11(a), (b) A four-color ordering for the stencil shown in Fig. 2.11(b)

in Eq. (2.23) are not diagonal matrices, then the unknowns of the same color can not be updated simultaneously. For the stencil in Fig. 2.11(a), three colors are needed to separate all unknowns into three groups, say the red, black and green unknowns, as shown in Fig. 2.12(a). Note that for the points on the boundary of the domain, either two layers of ghost points or a lower order finiter difference scheme should be used to discretize the given PDE. For the stencil in Fig. 2.11(b), four colors are needed to separate all unknowns into four groups, say the red, black, yellow and green unknowns, so that all unknowns of the same color can be updated simultaneously. Fig. 2.12(b) shows the color pattern for a two-dimensional domain. In this case , a complete relaxation iteration includes four steps, in each of which the unknowns of a specific color are updated in parallel.

2.3 Parallel ADI Algorithm

The Alternating Direction Implicit (ADI) method was first used by Peaceman and Rachford for solving parabolic PDEs [148, 149] in 1950s. Since then, it has been widely used in many applications. The detailed discussions about this method can be found in [183]. Implementations of the ADI method on different parallel computers are discussed in [27, 101, 144, 156, 159].

2.3.1 ADI Algorithm

The solution process of the ADI algorithm can be best explained using the model equation

$$\frac{\partial u}{\partial t} = \frac{\partial^2 u}{\partial x^2} + \frac{\partial^2 u}{\partial y^2}, \qquad (x,y,t) \in \Omega \times [0,T], \qquad (2.31)$$

$$u(x,y,t) = 0, \qquad (x,y) \in \partial\Omega, \qquad t > 0,$$

$$u(x,y,0) = u_0(x,y), \qquad (x,y) \in \Omega,$$

where Ω is a unit square as shown in Fig. 2.2, $\partial\Omega$ is the boundary of Ω. If the central finite difference scheme with a five-point stencil is used to discretize the spatial derivatives in Eq.(2.31), we will have, at point (i,j), the following equation:

$$\frac{\partial u_{i,j}}{\partial t} = \frac{u_{i-1,j} - 2u_{ij} + u_{i+1,j}}{h^2} + \frac{u_{i,j-1} - 2u_{ij} + u_{i,j+1}}{h^2}, \qquad (2.32)$$

$$i = 1, 2, \cdots NI, \quad j = 1, 2, \cdots NJ.$$

Instead of treating both terms in the right hand side of Eq. (2.32) implicitly, we can treat one term implicitly and the other term explicitly, which gives rise to the following equations

$$\frac{u_{i,j}^{n+\frac{1}{2}} - u_{i,j}^n}{\frac{1}{2}\Delta t} = \frac{u_{i-1,j}^{n+\frac{1}{2}} - 2u_{i,j}^{n+\frac{1}{2}} + u_{i+1,j}^{n+\frac{1}{2}}}{h^2} + \frac{u_{i,j-1}^n - 2u_{i,j}^n + u_{i,j+1}^n}{h^2}, \tag{2.33}$$

$$i = 1, 2, \cdots NI, \quad j = 1, 2, \cdots NJ,$$

and

$$\frac{u_{i,j}^{n+1} - u_{i,j}^{n+\frac{1}{2}}}{\frac{1}{2}\Delta t} = \frac{u_{i-1,j}^{n+\frac{1}{2}} - 2u_{i,j}^{n+\frac{1}{2}} + u_{i+1,j}^{n+\frac{1}{2}}}{h^2} + \frac{u_{i,j-1}^{n+1} - 2u_{i,j}^{n+1} + u_{i,j+1}^{n+1}}{h^2}, \tag{2.34}$$

$$i = 1, 2, \cdots NI, \quad j = 1, 2, \cdots NJ.$$

Fig. 2.13(a) and (b) give an intuitive explanation of Eqs. (2.33) and (2.34). In Fig.

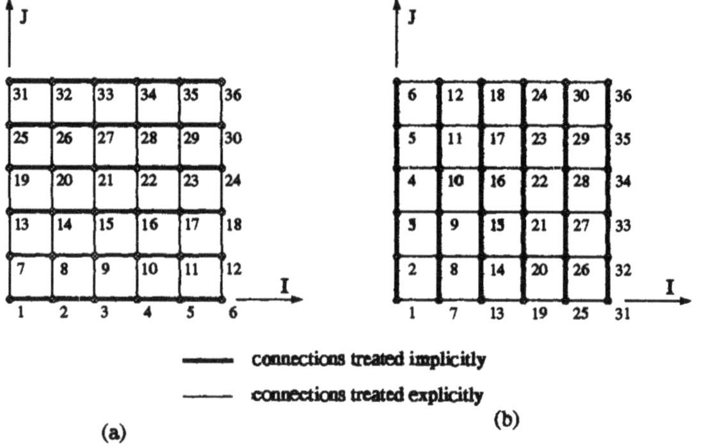

Figure 2.13: (a) All connections in the I direction are treated implicitly
(b) All connections in the J direction are treated implicitly

2.13(a), all connections in the I direction for the grid points are treated implicitly, while the connections in the J direction are treated explicitly, this corresponds to Eq. (2.33). If we assemble all equations in Eq. (2.33) and consider the explicit relation between equations on different J lines, we will have

$$(I - \frac{\Delta t}{2}A)u_{\bullet,j}^{n+\frac{1}{2}} = Eu_{\bullet,j-1}^n + Du_{\bullet,j}^n + Fu_{\bullet,j+1}^n \tag{2.35}$$

$$j = 1, 2, \cdots NJ,$$

where

$$u^n_{*,j} = \{u^n_{1,j}, u^n_{2,j}, \cdots, u^n_{NI,j}\}^T, \qquad j = 1, 2, \cdots NJ,$$

$$u^n_{*,0} = u^n_{*,NJ+1} = 0,$$

$$u^n_{0,j} = u^n_{NI+1,j} = 0,$$

$$A = \frac{1}{h^2}
\begin{bmatrix}
-2 & 1 & & \\
1 & \ddots & \ddots & \\
& \ddots & \ddots & 1 \\
& & 1 & -2
\end{bmatrix}
\underbrace{\qquad\qquad}_{NI},$$

$$D =
\begin{bmatrix}
1 - \frac{\Delta t}{h^2} & & \\
& \ddots & \\
& & 1 - \frac{\Delta t}{h^2}
\end{bmatrix},$$

$$E = F =
\begin{bmatrix}
\frac{\Delta t}{2h^2} & & \\
& \ddots & \\
& & \frac{\Delta t}{2h^2}
\end{bmatrix}.$$

There are actually NJ independent equation systems in Eq. (2.35) with tri-diagonal coefficient matrices. Instead of solving $NI \times NJ$ coupled linear equations with a penta-diagonal coefficient matrix as given in Eq. (2.7), we now solve NJ independent equation systems each of which has NI equations with a tri-diagonal coefficient matrix. A straightforward parallel version of this algorithm will be to assign one system in Eq. (2.35) to a processor so that different processors can work on different tri-diagonal systems in Eq. (2.35) simultaneously.

After finishing the first step described by Eq. (2.33), we need to continue to the second step given by Eq. (2.34). If we still use the same lexicographic ordering as was used for solving Eq. (2.35), then the assembled matrix equation system corresponding to Eq. (2.34) is

$$(I - \frac{\Delta t}{2}B)u^{n+1} = (I + \frac{\Delta t}{2}A)u^{n+\frac{1}{2}} \tag{2.36}$$

where

$$B = \frac{1}{h^2}
\begin{bmatrix}
-2I & I & & \\
I & -2I & \ddots & \\
& \ddots & \ddots & I \\
& & I & -2I
\end{bmatrix},$$

I is a $NI \times NI$ identity matrix, and

$$A = \begin{bmatrix} A_1 & & & \\ & A_2 & & \\ & & \ddots & \\ & & & A_{NJ} \end{bmatrix},$$

$$A_j = \frac{1}{h^2} \underbrace{\begin{bmatrix} -2 & 1 & & \\ 1 & -2 & 1 & \\ & & \ddots & 1 \\ & & 1 & -2 \end{bmatrix}}_{NI},$$

$$j = 1, 2, \cdots NJ.$$

Although the coefficient matrix in Eq. (2.36) is tri-diagonal, the bandwidth is $2 *$ $NI - 1$, and all $NI \times NJ$ equations are coupled together. Therefore, it is not suitable for parallel processing. If we reorder the points as shown in Fig. 2.13(b), i.e. ordering the points in the J direction first, followed by the points in the I direction, then we will have

$$(I - \frac{\Delta t}{2} A) u_{i,*}^{n+1} = E u_{i-1,*}^{n+\frac{1}{2}} + D u_{i,*}^{n+\frac{1}{2}} + F u_{i+1,*}^{n+\frac{1}{2}}, \tag{2.37}$$

$$i = 1, 2, \cdots NI,$$

where

$$u_{i,*} = \{u_{i,1}, u_{i,2}, \cdots, u_{i,NJ}\}^T, \qquad i = 1, 2, \cdots NI,$$

$$u_{0,*} = u_{NI+1,*} = 0,$$

$$u_{i,0} = u_{i,NJ+1} = 0.$$

The matrices A, E, D, and F have exactly the same forms as in Eq. (2.35), except that the order of matrices is $NJ \times NJ$, rather than $NI \times NI$.

A complete ADI step can now be given as:

Algorithm 2.3
ADI Algorithm

1. solve the NJ independent linear equation systems with tri-diagonal coefficient matrices in Eq. (2.35) to get $u_{*,j}^{n+\frac{1}{2}}$, $j = 1, 2, \cdots NJ$.
2. reorder the grid points.
3. solve the NI independent linear equation systems with tri-diagonal coefficient matrices in Eq. (2.37) to get $u_{i,*}^{n+1}$, $i = 1, 2, \cdots NI$.

For more general PDEs like

$$\frac{\partial u}{\partial t} = \frac{\partial}{\partial x}(b_1(x,y)\frac{\partial u}{\partial x}) + \frac{\partial}{\partial y}(b_2(x,y)\frac{\partial u}{\partial y}), \tag{2.38}$$

we will still have similar equations like those given by Eqs. (2.35) and (2.37). However, the elements in matrices A, D, E, and F will no longer be the same constants.

2.3.2 ADI Algorithm on a Ring of Parallel Processors

As shown by Eqs. (2.35) and (2.37), a complete step of the ADI method calls for the solution of two sets of independent equation systems. There are NI and NJ independent linear algebraic systems in Eqs. (2.35) and (2.37), respectively. A straightforward implementation of this algorithm on parallel computers would be, assuming the number of processors P is smaller than NI and NJ, to assign different tri-diagonal equation systems to different processors, so that $u_{*,j}^{n+\frac{1}{2}}$, $j = 1, 2, \cdots NJ$, can be solved simultaneously. After $u_{*,j}^{n+\frac{1}{2}}$ has been solved, different processors can then work on different tri-diagonal systems in Eq. (2.37) to solve $u_{i,*}^{n+1}$, $i = 1, 2, \cdots NI$, simultaneously. On a shared-memory parallel computer, all processors have accesses to the common memory. Therefore, this implementation is straightforward and does not cause any data movement. The only problem is that if $u_{i,j}^{n}$ is stored in an array $U(i,j)$, then the data in $U(i,j)$ will be accessed row by row when solving Eq. (2.35) and column by column in Eq. (2.37). Depending on how an array is mapped into the common memory, one access pattern is faster than the other because of high speed caches. For example, access by columns is faster than that by rows in Fortran.

On distributed memory parallel computers, however, the situation is different.

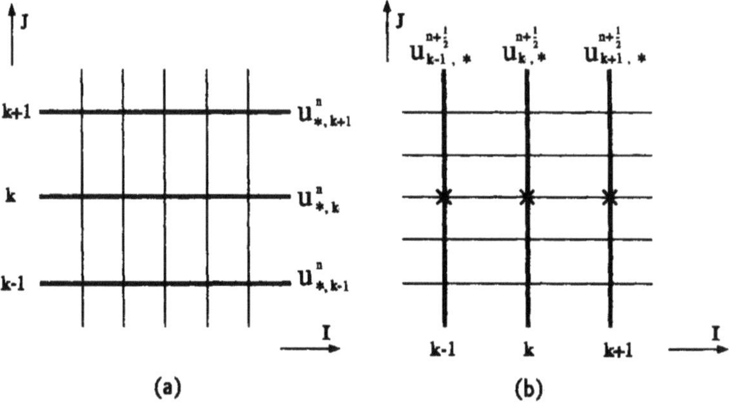

Figure 2.14: (a) Data availability requirement for solving the kth system in Eq. (2.35): $u^n_{*,k-1}$, $u^n_{*,k}$, $u^n_{*,k+1}$. (b) Data availability requirement for solving the kth system in Eq. (2.37): $u^{n+\frac{1}{2}}_{k-1,*}$, $u^{n+\frac{1}{2}}_{k,*}$ and $u^{n+\frac{1}{2}}_{k+1,*}$. After the first half-step, the data needed for the second half-step are available only at the points marked by $\times$ before the global message exchange.

Assume processor P_k is assigned to solve the kth tri-diagonal system in Eq. (2.35)

$$(I - \frac{\Delta t}{2}A)u_{*,k}^{n+\frac{1}{2}} = Eu_{*,k-1}^{n} + Du_{*,k}^{n} + Fu_{*,k+1}^{n} \tag{2.39}$$

in the first half-step of the ADI method, and that the solution values from the previous time steps at $j = k - 1,\ k,\ k + 1$ are available in its local memory as shown in Fig. 2.14(a), then Eq. (2.39) can be solved to obtain $u_{*,k}^{n+\frac{1}{2}}$. At this point, processor P_k has $u_{*,k}^{n+\frac{1}{2}}$ in its local memory. Now, in the second half-step of the ADI method, if

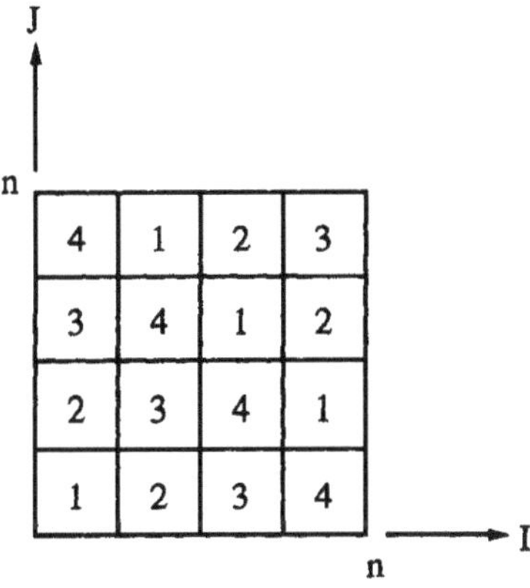

Figure 2.15: Assignment of subdomains to 4 processors. The original domain is divided into subdomains. The number inside each subdomain represents the processor to which the subdomain is assigned.

processor P_k is assigned to solve the kth system in Eq. (2.37)

$$(I - \frac{\Delta t}{2}A)u_{k,*}^{n+1} = Eu_{k-1,*}^{n+\frac{1}{2}} + Du_{k,*}^{n+\frac{1}{2}} + Fu_{k+1,*}^{n+\frac{1}{2}}, \tag{2.40}$$

then P_k must have $u_{k-1,*}^{n+\frac{1}{2}}$, $u_{k,*}^{n+\frac{1}{2}}$ and $u_{k+1,*}^{n+\frac{1}{2}}$ in its local memory as shown in Fig. 2.14(b). However, after the first half-step, $u_{k-1,*}^{n+\frac{1}{2}}$, $u_{k,*}^{n+\frac{1}{2}}$ and $u_{k+1,*}^{n+\frac{1}{2}}$ are held by different processors. Therefore, a global message exchange (each processor P_i communicates with all other processors) is needed to get all necessary data before Eq. (2.40) can be solved. After Eq. (2.40) has been solved, Processor P_k will have $u_{k,*}^{n+1}$ in its local memory. A similar data exchange is required to prepare each processor for the first half-step of the ADI algorithm at the next time level. Therefore, two global message exchanges are needed at each ADI step if this straightforward implementation is used. Each global data exchange is similar to transposing a matrix so that the matrix is distributed to different processors by rows and columns alternately. Since each processor has to communicate with all other processors in this global message exchange, it could be very time-consuming.

One way to avoid global data exchange on a ring of processors is to distribute data and equations as shown in Fig. 2.15 where a two-dimensional domain is decomposed into subdomains and assigned to $k = 4$ processors. The number inside a subdomain indicates the processor number to which the subdomain is assigned. By assigning a subdomain to a processor, we mean the availability of all data needed to generate difference equations at the points inside the subdomain, and the solution values of the previous time step at all points inside the subdomain as well as the ghost points around the subdomain. Assume the domain in Fig. 2.15 is $n \times n$ with n divisible by the number of processors k. In the first half-step, each tri-diagonal equation system in Eq. (2.35) is split into k equal parts and assigned to k processors. The tri-diagonal equation system can be solved by a forward and backward elimination processes:

$$
\begin{bmatrix}
a_1 & b_2 & & \\
c_1 & a_2 & \ddots & \\
& \ddots & \ddots & b_n \\
& & c_{n-1} & a_n
\end{bmatrix}
\begin{bmatrix}
x_1 \\ x_2 \\ \vdots \\ x_n
\end{bmatrix}
=
\begin{bmatrix}
f_1 \\ f_2 \\ \vdots \\ f_n
\end{bmatrix}
\qquad
\begin{array}{c} \text{step 1} \\ \longrightarrow \end{array}
$$

$$
\begin{bmatrix}
a_1' & b_2' & & \\
& a_2' & \ddots & \\
& & \ddots & b_n' \\
& & & a_n'
\end{bmatrix}
\begin{bmatrix}
x_1 \\ x_2 \\ \vdots \\ x_n
\end{bmatrix}
=
\begin{bmatrix}
f_1' \\ f_2' \\ \vdots \\ f_n'
\end{bmatrix}
\qquad
\begin{array}{c} \text{step 2} \\ \longrightarrow \end{array}
$$

$$
x_n = \frac{f_n'}{a_n'}, \quad x_i = (f_i' - b_{i+1}'x_{i+1})/a_i', \quad i = n-1, \cdots 1.
$$

The straightforward implementation of the elimination process is sequential. The forward elimination starts from the processor holding the first part of the tri-diagonal system, and continues to the other processors. Once a processor finishes elimination of its part, it must pass the last equation to its succeeding processor so that the elimination can continue. The backward elimination starts from the processor holding the last part of the original equation. Once a processor finishes the backward elimination, it passes the last x_i solved to its preceding processor to continue the elimination process. With one tri-diagonal equation system, only one processor works at a given time while the other $k - 1$ processors stay idle. If there are k tri-diagonal equation systems and different processors take different parts in different equation systems, then all processor will stay busy eliminating different parts in different equation systems. As shown in Fig. 2.15, there are totally n tri-diagonal equation systems and $k = 4$ processors, we can divide n systems into k groups. Each group consists of n/k systems. At the beginning, each processor works on the first part of its group of equation systems. Upon finishing elimination of the first part, each processor passes the pivoting equation to the succeeding processor and receives a pivoting equation from the preceding processor, then all processors start working on the second part in different groups of equation systems. For example, processor 1 starts from the first part in the first group of equation systems. When finished, it passes the last equation to processor 2 so that processor 2 can work on the second part in the first group. It then receives th pivoting equation from processor 4 and starts working on the second part of the 4th group. The process continues until it finishes eliminations of the third part in the third group and the last part in the second group. In the backward elimination, the progress of elimination is in a reversed direction — from right to left in Fig. 2.15.

In the second half-step, each equation system in Eq. (2.37) is also divided into k parts and assigned to k different processors. On processor 1, data are available to form the equations for the first, second, third and the last parts of group 1, 4, 3 and 2 respectively. Once all these parts have been formed, similar forward and backward processes can progress from bottom-up and top-down respectively. No global data exchange is needed. Only local data exchange (between the predecessor and successor) is necessary to continue the elimination processes. Fig. 2.16 shows the progress of the solution process in each step. There are totally $k = 4$ equation groups G_1, G_2, G_3 and G_4, each of which contains n/k tri-diagonal equation systems. Each one of the tri-diagonal systems is further divided into $k = 4$ parts which are assigned to k different processors. For the forward elimination part, processor 1 works on part 1 of equation group G_1 in time period t_1, part 2 of group G_4 in period t_2, part 3 of G_3 in period t_3, and part 4 of G_2 in period t_4. The progresses for all other processors can be read from the rest rows in Fig. 2.16. The backward elimination process progresses in the opposite direction.

2.3.3 ADI Algorithm on a 2-D Grid of Parallel Processors

In this case, we assume that there are k^2 processors arranged as a $k \times k$ grid. The domain in which solutions are sought is $n \times n$. Again for simplicity, we assume

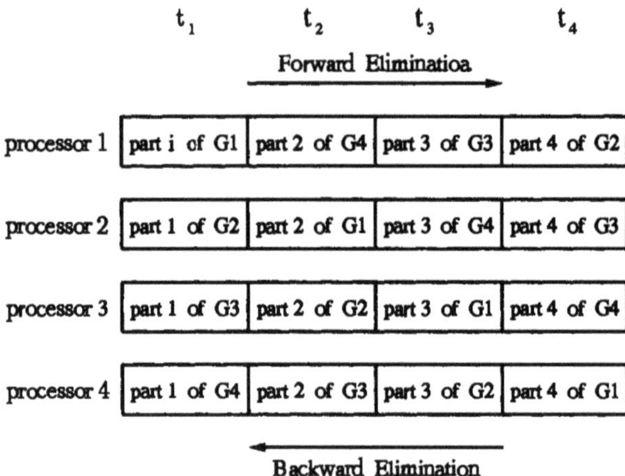

Figure 2.16: Progresses of the forward and backward eliminations on different processors

n is divisible by k. The domain is decomposed into subdomains and assigned to processors as shown in Fig. 2.17. Although Fig. 2.17 looks similar to Fig. 2.15, the assignments of subdomains to processors are totally different. In Fig. 2.15, each processor is assigned 4 subdomains, while in Fig. 2.17 each processor is assigned only one subdomain. In each step, the tri-diagonal equation systems in Eq. (2.35) and Eq. (2.37) are divided into different groups (4 groups in this case). Each single tri-diagonal equation system is further divided into 4 equal parts and assigned to different processors. In the first half-step, processors 1, 2, 3 and 4 will be assigned the first group of equation systems in Eq. (2.35), while in the second half-step, processors 1, 5, 9 and 13 will be assigned the first group of equation systems in Eq. (2.37). The assignments of equation groups to different processors at different steps are shown in Fig. 2.17.

In each equation group, there are n/k tri-diagonal equation systems. The ith equation group includes equation systems from Eq. (2.35) or Eq. (2.37) with index j ranging from $(i-1)\frac{n}{k}+1$ to $\frac{in}{k}$. When a processor finishes the elimination process for a particular part of a particular equation system, it passes the pivoting equation to the succeeding processor so that the succeeding processor can work on the next part of the same equation system. For example, as shown in Fig. 2.18, in the forward elimination

	Group 1	Group 2	Group 3	Group 4
Group 4	13	14	15	16
Group 3	9	10	11	12
Group 2	5	6	7	8
Group 1	1	2	3	4

Figure 2.17: A two-dimensional domain is decomposed into subdomains and assigned to a grid of $4 \times 4 = 16$ processors

of the first half-step, processor 1 first starts working on the elimination of the first part of equation system 1 in Group 1, which is the equation system in Eq. (2.35) with $j = 1$, while the other three processors 2, 3 and 4 stay ideal. Once processor 1 finishes elimination of the first part of equation system 1, it passes the pivoting equation to processor 2 so that processor 2 can work on the second part of equation system 1. Processor 1 then starts working on the first part of the second equation system with $j = 2$ in Eq. (2.35). The process continues as a pipeline. After three stages, all four processors will be busy working on different parts of different equation systems. For this algorithm to work, it is assumed that the number of equation systems in a group must be larger than the number of processors assigned to the equation group. The elimination process for other groups of equation systems assigned to processors 5–16 are the same. The only difference is that equation systems in different groups have different j indices from Eq. (2.35). The backward elimination proceeds in exactly the same way, except in an opposite direction.

In the second half-step, the assignment of equation groups to processors should be changed as indicated by columns in Fig. 2.17, i.e. processors 1, 5, 9, 13 are assigned equation group 1 in Eq. (2.37) with $1 \leq i \leq n/k$, processors 2, 6, 10, 14 are assigned equation group 2 in Eq. (2.37) with $n/k + 1 \leq i \leq 2n/k$, where $k = 4$. The other two equation groups are assigned to processors 3, 7, 11, 15 and processors 4, 8, 12, 16, respectively. The pipeline for the forward elimination goes bottom-up, while the pipeline for the backward elimination goes top-down.

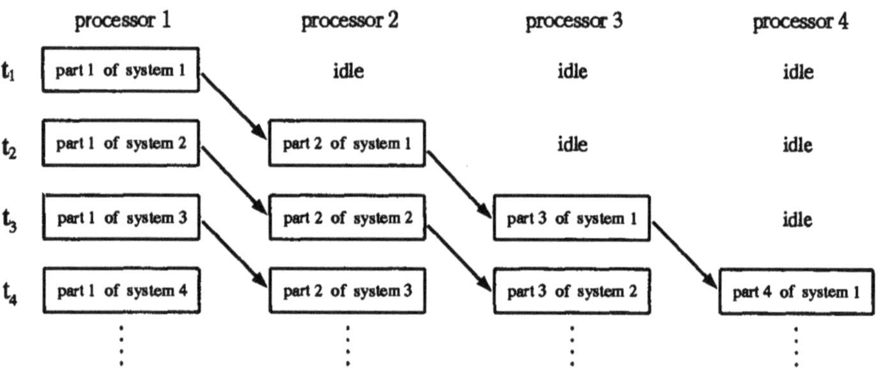

Figure 2.18: progress of pipelined elimination process. As time t_i goes on, more and more processors become active.

2.3.4 Further Improvement on Parallel ADI Algorithm

The parallel ADI algorithms discussed in the previous two sections are all based on solving different tri-diagonal equation systems concurrently by different processors. Each tri-diagonal equation system is still solved sequentially by the standard elimination method. These algorithms can be further improved by parallelizing the solution processes of all tri-diagonal equation systems.

As discussed before, the major computation in the ADI algorithm is the solution of the tri-diagonal equation systems given by Eq. (2.35) and Eq. (2.37). The process can be parallelized by several methods, like the cyclic reduction discussed in section 1.3.1, or the incomplete nested dissection [64, 184] to be discussed below.

Assume that the $n \times n$ tri-diagonal equation system

$$Tx = b \qquad (2.41)$$

is to be solved by k processors, n/k is an integer, and each processor is assigned n/k consecutive equations of (2.41). Then, processor j will take equations numbered from $(j-1)\frac{n}{k}+1$ to $j\frac{n}{k}$. Fig. 2.19 shows the initial distribution of rows (equations) of T to different processors.

The parallel solution process of Eq. (2.41) consists of three steps:

step1: The first step is the simultaneous forward and backward eliminations by all processors. In the forward elimination, on each processor j, the subdiagonal

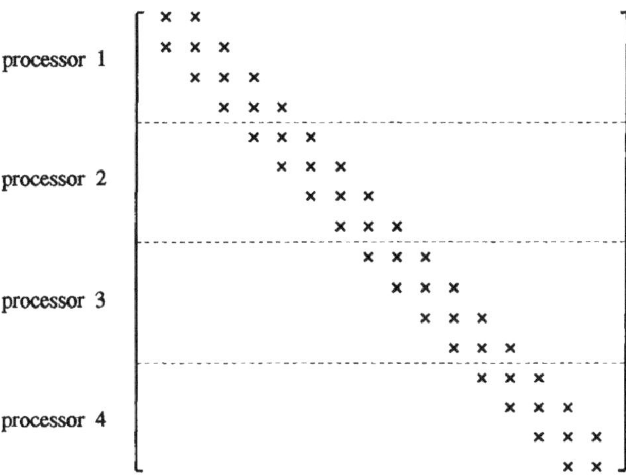

Figure 2.19: Initial distribution of rows of T which is a 16×16 tri-diagonal matrix to 4 processors.

elements of row $\frac{n(j-1)}{k} + 2$ through row $\frac{nj}{k}$ are eliminated. These eliminations on different processors are completely independent and require no inter-processor communication. At the end of the forward elimination, the equation system will look like the one shown in Fig. 2.20. In the backward elimination process, going bottom-up, processor j, $j = 2, \cdots k - 1$, eliminates the superdiagonal elements of equations numbered $\frac{nj}{k} - 2$ to $\frac{n(j-1)}{k}$. Processor 1 eliminates superdiagonal elements for equations $\frac{n}{k} - 2$ to 1, and processor k eliminates superdiagonal elements for equations $n - 1$ to $\frac{(k-1)n}{k}$. After the backward elimination process, the matrix will look like the one shown in Fig. 2.21. Since in this process the superdiagonal element in the last equation of processor j, $j = 1, 2, \cdots k - 1$, must be eliminated by using the first equation in processor $j + 1$, interprocessor communication is needed for processor j to get the updated row $j\frac{n}{k} + 1$.

step2: Fig. 2.21 shows that the unknowns $x_{\frac{nj}{k}}$, $j = 1, 2, \cdots k - 1$, in Eq. (2.41) satisfy another set of $k - 1$ tri-diagonal equations as indicated by the rows in Fig. 2.21 marked by $\star$, which involve x_4, x_8 and x_{12}. The $k - 1$ tri-diagonal equations are distributed to the $k - 1$ processors. The solution process of these equations by the $k - 1$ processors is largely sequential. However, since each

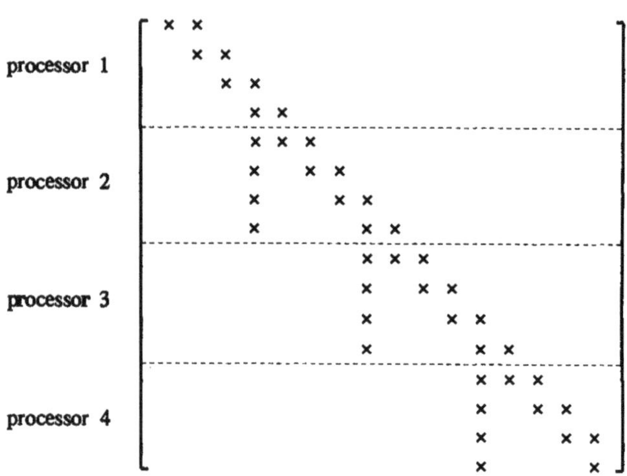

Figure 2.20: Matrix structure after the forward elimination

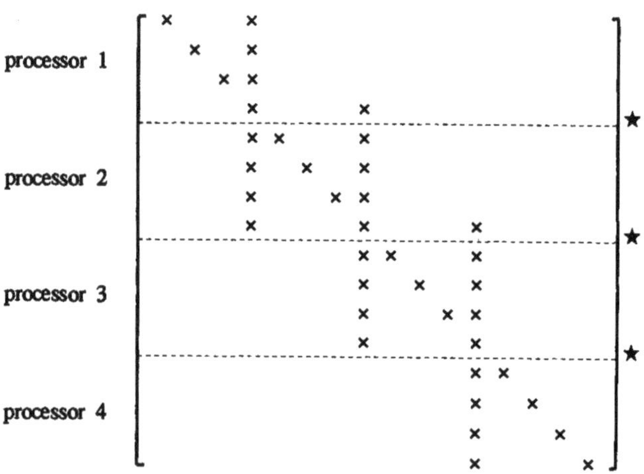

Figure 2.21: Matrix structure after the backward elimination

processor has only one equation, the solution complexity is independent of the problem size. The major work of the second step is to solve the $k-1$ tri-diagonal equations, say by the regular Gaussian elimination.

step3: Once the unknowns $x_{\frac{nj}{k}}$, $j = 1, 2, \cdots k - 1$, have been solved, it is clear from Fig. 2.21 that each processor can solve for the other unknowns by subtracting multiples of the fill-in columns from the equations it holds. The computations on different processors can be done in parallel. The only communication step is that the unknown $x_{\frac{nj}{k}}$ calculated at step 2 by processor j, $j = 1, 2, \cdots k - 1$, has to be passed to processor $j + 1$ before the solution process for the rest of unknowns can start.

With the parallelized tri-diagonal equation solver, it is possible to finish an ADI iteration with execution time $O(n^{\frac{2}{3}})$ which is better than $O(n)$ if sequential tri-diagonal equation solver is used [100].

2.3.5 Three-Dimensional Problems

For problems defined in three-dimensional domains, the basic idea presented in the previous sections is still applicable. In addition to Eqs. (2.35) and (2.37), there is another set of equation systems in the third dimension. As shown in Fig. 2.22, for a three-dimensional $n \times n \times n$ domain which has been decomposed into columns and assigned to different processors, the processing procedures in the x direction (index I) and y direction (index J) are exactly the same as discussed in section 2.3.3, except in each direction a processor will be handling a section of $\frac{n^2}{k}$ tri-diagonal systems, rather than $\frac{n}{k}$ systems as in the two-dimension case. For the third dimension, each processor solves $(\frac{n}{k})^2$ independent tri-diagonal systems of n equation each. The processes on all processor are perfectly parallel.

2.4 Parallel Multigrid Method

2.4.1 Multigrid Method

The multigrid method [20, 22, 75, 76] is very efficient in accelerating the convergence of many simple iterative algorithms, especially when applied to elliptic and parabolic equations. The basic idea is to use multiple grids to eliminate error components of different frequencies. As discussed in section 2.2, many iterative algorithms can be written in the form of

$$u_k = Tu_{k-1} + c, \qquad k = 1, 2, \cdots, \tag{2.42}$$

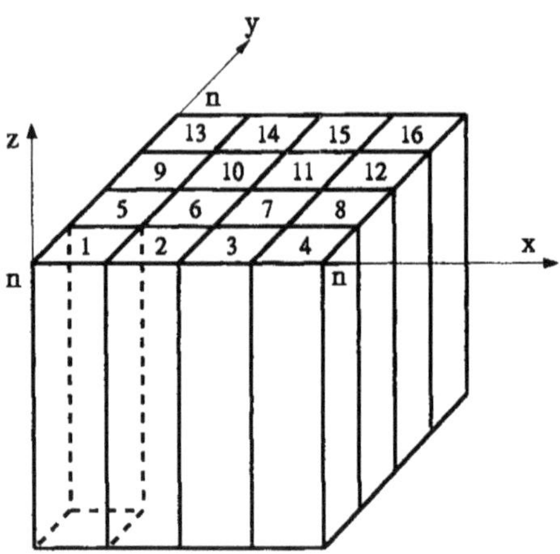

Figure 2.22: A three-dimensional domain decomposed into columns. The numbers on top of the columns indicate the processor numbers.

where u_0 is the initial guess used to start the iterations. Assume u^* is the true solution of (2.42), so that

$$u^* = Tu^* + c. \tag{2.43}$$

Subtracting (2.43) from (2.42), we obtain

$$u_k - u^* = T(u_{k-1} - u^*).$$

Let $e_k = u_k - u^*$, which is the error vector between the true solution and the approximate solution after k iterations, we have

$$||e_k|| = ||Te_{k-1}|| \leq ||T|| \cdot ||e_{k-1}|| \leq \cdots \leq ||T||^k ||e_0||. \tag{2.44}$$

Expression (2.44) shows that the relaxation process is guaranteed to converge if $||T|| < 1$. The smaller the $||T||$, the faster the convergence rate. It also shows that a relaxation process is actually a process to gradually reduce the norm of an error vector. Note that an arbitrary initial error vector $e_0 \in R^n$ can be represented as a linear combination of Fourier modes with different frequencies, we then have two observations that naturally lead to the multigrid method.

The first is that it has been observed that on a given grid, the error components of different frequencies (modes) are reduced at different rates by the relaxation processes.

As an example, assume that the initial error e_0 is the superposition of the two curves shown in Fig. 2.23(a) which is a smooth component of low frequency and Fig. 2.23(b)

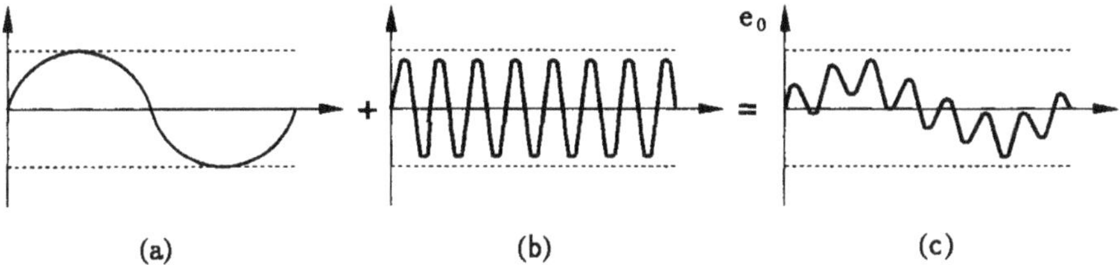

Figure 2.23: (a) A smooth error component of low frequency. (b)An oscillating error component of high frequency. (c) The initial error as a result of superposition of the components in (a) and (b).

which is a more oscillating component of high frequency. After several iterations, the error e_k will look like the curve shown in Fig. 2.24. The amplitude of the remaining

Figure 2.24: The smoothed error e_k after several relaxations. The oscillating component has been reduced significantly, but the smooth component still remains pretty much intact.

error will decrease very slowly as the relaxation process continues. It can be seen from Fig. 2.24 that the error e_k is now a smooth curve, which means that the oscillating component in the error has been very much reduced. Therefore, a relaxation scheme is also frequently referred to as a smoother. It is very efficient in damping out high frequency error components on a given grid, but can not do much about the smooth error components.

The second observation is that what appears to be a smooth error on a fine grid may look more oscillatory on a coarse grid when the error is projected from the fine grid to a coarse grid. Therefore, the concepts of smoothness and oscillation are grid dependent, as shown in Fig. 2.25. Basically, oscillation indicates the rate of change

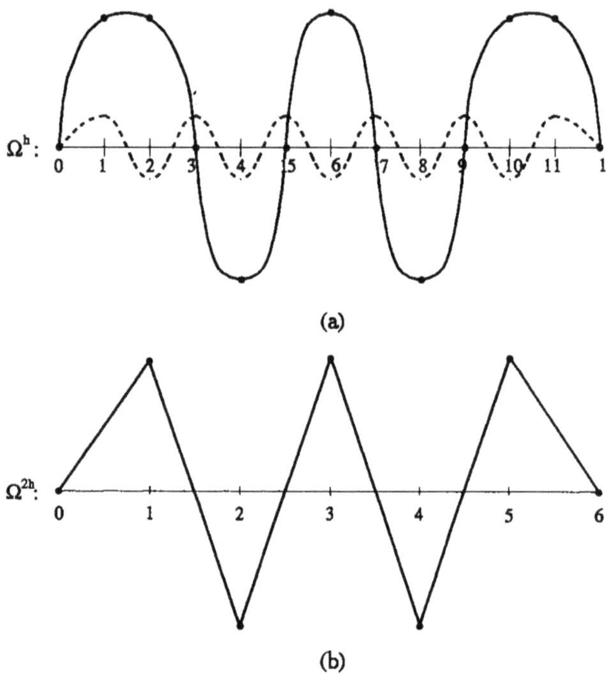

Figure 2.25: (a) A smooth error on a fine grid with 13 grid points (b) The same error projected from the fine grid to a coarse grid with only 7 grid points.

of errors. On a grid with finite number of grid points, the change of Fourier modes is characterized by the wave number which is the number of humps and dips on the curve. For Fig. 2.25(a), with 13 grid points, the most oscillatory error representable is the one with a wave number of 11, which is the curve represented by the dashed line. Therefore, the curve with a wave number of 5 is not the most oscillatory error on that grid. However, when this error is projected to a coarse grid with only 7 grid points, it is the error with the maximum number of waves representable on this coarse grid. Therefore, it is the most oscillatory error representable on that coarse grid.

Based on these two observations, one would expect that the smooth error component which can not be reduced efficiently by iterations on a fine grid should be reduced much faster if the iterations are carried out on a coarse grid. This is exactly the key step used in the multigrid method to improve the efficiency of basic iterative algorithms, like the Jacobi, Gauss-Seidel, and SOR algorithms.

To facilitate the discussion, we introduce the concepts of residual and residual equation. Assume that after discretization the equation system we want to solve is

$$Ax = b, \tag{2.45}$$

the residual is defined as the vector r with

$$r = b - Ax', \tag{2.46}$$

where x' is an approximate solution of Eq. (2.45). If the error is defined as

$$e = x - x',$$

it is easy from Eqs. (2.45) and (2.46) to obtain the residual equation

$$Ae = r, \tag{2.47}$$

which shows that the error vector e satisfies the original equation system with the residual vector as the right hand side term. The true solution (or a better solution if e is not solved exactly) can then be represented as

$$x = x' + e. \tag{2.48}$$

Both e and r are measures of how close the approximate solution x' is to the true solution. The vector e is a better measure of the error, but is just as inaccessible as the exact solution. The residual vector is computable with an approximate solution, but is less reliable since a small norm of residual vector does not always mean a small error in the solution.

If the basic relaxation algorithms are applied to Eq. (2.45) to obtain an approximate solution x' on a given grid, it can be expected that the convergence rate of the iteration will slow down and eventually stall once the oscillatory error components have been eliminated. At this point, the vectors e and r contain only smooth components which can be represented well on a coarse grid. Once these vectors are projected to the coarse grid, they look more oscillatory than they do on the fine grid. Therefore, it is natural to project vector r to a coarse grid and solve Eq. (2.47) on a coarse grid to obtain the correction vector e. This correction vector is then interpolated back to the fine grid to obtain a better solution using Eq. (2.48). This process can be described as:

Algorithm 2.4
Two-grid Algorithm

Iterate on $Ax = b$ on Ω^h to get an approximation x^h.

Compute the residual $r^h = b - Ax^h$.

Iterate on $A^{2h}e^{2h} = r^{2h}$ on Ω^{2h} to obtain an approximation e^{2h}.

Correct the approximation x' on Ω^h by
$$x^h \longleftarrow x^h + e^{2h}.$$

---

where Ω^h and Ω^{2h} represent the fine and coarse grids respectively. Note in the above algorithm, we need to specify how A^{2h} is defined and how the vectors r and e are transferred between grids. The projection operator I_h^{2h} which projects a vector from a fine grid to a coarse grid can be defined as a simple injection

$$I_h^{2h}x^h = x^{2h} \tag{2.49}$$

where

$$x_i^{2h} = x_{2i}^h, \quad i = 0, 1, \cdots, \frac{N}{2},$$

or as a full weighting operator

$$x_i^{2h} = \frac{1}{4}(x_{2i-1}^h + 2x_{2i}^h + x_{2i+1}^h), \quad i = 0, 1, \cdots \frac{N}{2},$$

with

$$x_{-1}^h = 0, \qquad x_{N+1}^h = 0.$$

The interpolation operator I_{2h}^h which prolongates a vector x^{2h} from a coarse grid to a fine grid can be defined as a linear interpolation operator

$$I_{2h}^h x^{2h} = x^h \tag{2.50}$$

where

$$x_{2i}^h = x_i^{2h}, \qquad\qquad i = 0, 1, \cdots, \frac{N}{2},$$

$$x_{2i+1}^h = \frac{1}{2}(x_i^{2h} + x_{i+1}^{2h}), \qquad i = 0, 1, \cdots, \frac{N}{2} - 1.$$

There are other forms for operators I_{2h}^h and I_h^{2h}. Readers are referred to reference [22] for matrix forms of I_{2h}^h and I_h^{2h} in one and two-dimensional grids. It is also shown in [22] that the operator A^{2h} is defined as

$$A^{2h} = I_h^{2h} A^h I_{2h}^h. \tag{2.51}$$

With definitions (2.49), (2.50) and (2.51), it can be seen that all transformations in the previous two-grid algorithm are well defined. Note that the iterations on Eq. (2.47) defined on grid Ω^{2h} will encounter the same convergence problem as that on Eq. (2.45) defined on Ω^h, thus another coarser grid Ω^{4h} can be used to further improve the convergence. Therefore, a multigrid algorithm can be defined recursively on more than two grids. The popular V-cycle multigrid algorithm is given by [22]:

Algorithm 2.5
***V*-cycle scheme $MV^h(x^h, b^h)$**

if (Ω^h is the coarsest grid) then
    solve $A^h x^h = b^h$ directly
else
    1. iterate n_1 times on $A^h u^h = b^h$ with x^h
       as the initial guess to get u^h
    2. $x^{2h} = 0$
    3. $b^{2h} = I_h^{2h}(b^h - Au^h)$
    4. $x^{2h} = MV^{2h}(x^{2h}, b^{2h})$
    5. $u^h = u^h + I_{2h}^h x^{2h}$
    6. iterate n_2 times on $Ax^h = b^h$ with u^h
       as the initial guess to get x^h
end if

In the above algorithm, the variable x represents both the solution to the original equation (2.45) (on the finest grid Ω^h) and the correction term from Eq. (2.47) (on coarse grids $\Omega^{2h}, \Omega^{4h}, \cdots$). Similarly, the variable b has been used to represent both the right hand side term of Eq. (2.45) and the residual vector in Eq. (2.47). Note in the above V-cycle algorithm, there are n_2 more iterations after the coarse grid correction term e^{2h} has been added to the approximate solution u^h on the fine grid. The progress of the V-cycle multigrid algorithm across different grids is shown in Fig. 2.26 where 6 grids are involved. The iteration for $A^h x^h = b^h$ starts on the finest grid

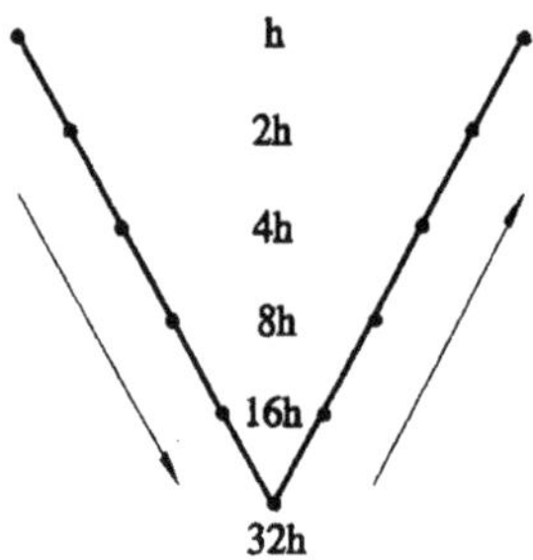

Figure 2.26: Progress of the V-cycle multigrid algorithm across different grids, h is the grid spacing on the finest grid. Each grid is visited twice.

Ω^h, then the residual $b^{2h} = I_h^{2h}(b^h - A^h u^h)$ is passed to the next coarse grid to solve $A^{2h} x^{2h} = b^{2h}$ on Ω^{2h}. After a few iterations, the residual $b^{4h} = I_{2h}^{4h}(b^{2h} - A^{2h} u^{2h})$ is then passed to grid Ω^{4h} to solve $A^{4h} x^{4h} = b^{4h}$. The process continues until it reaches the coarsest grid Ω^{32h} on which the system $A^{32h} x^{32h} = b^{32h}$ is usually solved using a direct method since the system should be small enough. This corresponds to the left half of a V-cycle in which the computation goes from the finest grid to the coarsest grid. Once the correction vector x^{32h} is solved on Ω^{32h}, it is prolongated to the next fine grid Ω^{16h} to compute a better solution $u^{16h} = u^{16h} + I_{16h}^{32h} x^{32h}$, followed by iterations on $A^{16h} x^{16h} = b^{16h}$ using u^{16h} as the initial guess to obtain x^{16h}. The process goes on until it finally gets back to the finest grid Ω^h, and x^h is calculated. Each grid is visited twice: One before and one after the coarse grid correction respectively. The restriction and prolongation operators defined in Eqs. (2.49) and (2.50) provide the mechanisms for the transfer of solutions and residuals between different grids.

The numbers of iterations n_1 and n_2 in the V-cycle algorithm are very small, usually 1 or 2. There are many different and more complicated multigrid schemes, like the W-cycle and F-cycle algorithms. Interested readers are referred to references [22, 75] for details.

2.4.2   Parallelization by Grid Splitting

Since the early papers by Grosch [70, 71] and Brandt [19], there has been a lot of
research on parallelizing the multigrid method for different architectures [30, 31, 84,
85, 126, 170, 171]. We will mainly discuss implementations on distributed memory
parallel computers. As discussed in the previous section, a complete V-cycle (and
similarly for other multigrid cycles) in the multigrid method involves iterations using
basic relaxation schemes on different grids, and transfer of data between different
grids by interpolation and prolongation. To parallelize the solution process, the original grid at the finest level can be split into subgrids which are then assigned to
different processors. As shown in Fig. 2.27, the original grid is split into four dif-

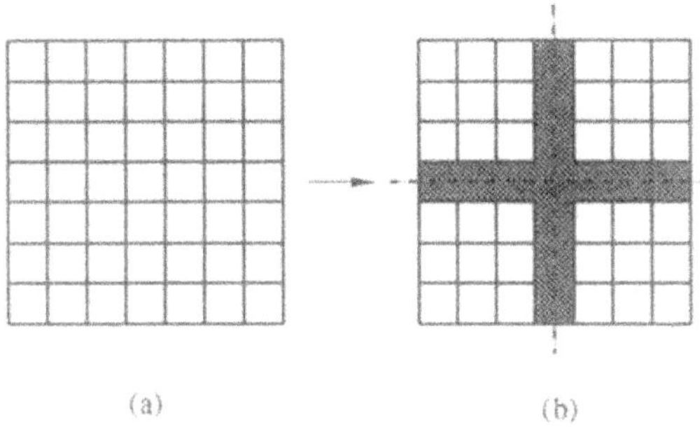

Figure 2.27: (a) The original grid, (b) The grid is split into four subgrids

ferent subdomains. If the Jacobi type relaxation algorithm is used, the relaxations
can be carried out in parallel on different processors without difficulty. If the Gauss-
Seidel and SOR type algorithms are used, the relaxation process can be carried out in
parallel by using the coloring schemes discussed in section 2.2.3.2. For the red-black
ordering, the order of processing for different colors is not important. However, for
multi-coloring with more than two colors, the order of processing for different colors
can affect the number of interprocessor communications [5]. For example, if the 9-
point stencil shown in Fig. 2.11(b) is used to discretize the relevant PDEs with the
four-color ordering, then the relaxations should be carried out in four steps on the
points of different colors separately. Fig. 2.28 shows that the grid is split into two
parts and assigned to processors P_A and P_B respectively. If the relaxations are carried out in the order of black→red→green→yellow, then after two processors finish
calculating new values for the black points, processor P_A must send new values at the

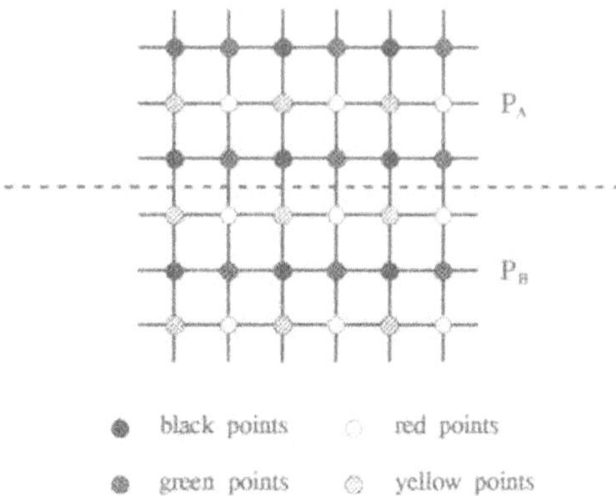

Figure 2.28: The four-color scheme on a two-dimensional grid. The upper part is assigned to processor P_A and the lower part is assigned to processor P_B.

black points on the boundary to processor P_B for updating values at the red points on the boundary. After the calculations on the red points have been finished, processor P_B must send the updated values at the red points on the boundary to processor P_A for updating values at the green points on the boundary. The situation is similar for relaxations on the points of other two colors. After each sweep for the grid points of a specific color, a message containing updated solution values at the grid points of that color must be sent from P_A to P_B, or vice versa. Therefore, each complete relaxation requires four message passings with each message containing solution values at half of the boundary points.

If the relaxation is carried out in the order of black→green→yellow→red, then there will be no need to pass solutions after the first relaxation on black points, since the relaxation on the green points for processor P_B does not need solutions at the black points on processor P_A. After the relaxation on the green points, processor P_A needs to send updated values of both the black and green points on the boundary to processor P_B for updating the yellow and red points on the boundary. The message contains solution values at all points on the boundary of the subdomain assigned to processor P_A. Similarly, processor P_B needs to send a message containing updated solutions at the yellow and red points on the boundary to processor P_A. Therefore, for each complete relaxation, two massage passings are needed. The message length is twice as long as that of the message in the previous relaxation scheme. Although the total amount of data transferred in each complete relaxation (four steps for four colors) is the same in both algorithms, the second algorithm is usually better since it requires only two communications in each relaxation.

2.4.3 Agglomeration

As discussed in the previous section, the original grid is split into smaller subgrids which are then assigned to different processors for parallel processing. The relaxations are carried out at all grid levels. At each level, message passing is needed to exchange solution values of the grid points on the boundaries of the subgrids. As the multigrid cycle goes down to coarse grids, the amount of computation decreases between each message passing. In the extreme case, it might happen that there are fewer grid points on the coarse grid than the number of available processors. Therefore, some processors will be idle waiting for other processors to complete the computation on coarse grids. To minimize the communication cost, a process called agglomeration [5, 60, 190] is often used to combine subgrids so that all processors will stay busy. The purpose of using agglomeration is to reduce the number of message exchange on coarse grids by duplicating computations on processors which would otherwise be idle.

2.4.4 Filtering Algorithm

The filtering algorithm was first used by Chan and Tuminaro [32, 173] to parallelize the multigrid cycles. The basic idea of the filtering algorithm is to create multiple correction problems which can be handled by processors that would otherwise be idle in the standard multigrid method. As discussed in Section 2.4.1, coarse grids are used in the multigrid method to eliminate low frequency errors in the solution. At a given grid level, after a few relaxations, the residual is projected on to a coarse grid for a new relaxation process to eliminate the low frequence errors more effectively. This is the so-called coarse grid correction process. It is clear from the discussions in the previous section that as the multigrid cycle proceeds to coarse grid correction, the amount of computation in each correction process decreases significantly. As a result, many processors may become idle.

With the filtering algorithm, one creates two correction problems, one on the coarse grid and the other on the fine grid. Instead of projecting the residual

$$r^h = b^h - A^h u^h \tag{2.52}$$

to the coarse grid to compute a correction for the approximate solution u^h, the vector r^h is split into two parts r_1^h and r_2^h which contain low and high frequency errors respectively. Correspondingly, the correction equation (2.47) is also split as

$$A^h e_1^h = r_1^h \qquad (2.53)$$

and

$$A^h e_2^h = r_2^h. \qquad (2.54)$$

The correction equation (2.53) is treated similarly as in the standard multigrid method: Both the operator A^h and the residual r_1^h that contains the low frequency error are projected to the coarse grid to form a new correction equation

$$A^{2h} e_1^{2h} = r_1^{2h}$$

where $A^{2h} = I_h^{2h} A I_{2h}^h$ and $r_1^{2h} = I_h^{2h} r_1$. The correction e_1^{2h} at the coarse grid is then prolongated back to the fine grid by

$$e_1^h = I_{2h}^h e_1^{2h}.$$

The correction equation (2.54) is solved on the fine grid (approximately) by relaxations since the residual r_2^h contains high frequency error. A better solution is then obtained by

$$u^h = u^h + e_1^h + e_2^h.$$

The calculation for solving Eqs. (2.53) and (2.54) can be done in parallel, and the two-level filtering algorithm discussed above can be defined recursively using the filtering procedure to solve the coarse grid correction problem (2.53) [173]:

Algorithm 2.6
Filter_MV^h(A^h, b^h, u^h, level)

if (level. eq. the coarsest grid level) then
    solve $A^h u^h = b^h$ exactly
else
    relax on $A^h u^h = b^h$
    calculate $r^h = (b^h - A^h u^h)$
    split r^h as $r^h = r_1^h + r_2^h$
    calculate $r^{2h} = I_h^{2h} r_1^h$
    Filter_$MV^{2h}(A^{2h}, r^{2h}, u_1^{2h}, \text{level} - 1)$
    relax on $A^h u_2^h = r_2^h$
    calculate $u_1^h = I_{2h}^h u_1^{2h}$
    update u^h by $u^h = u^h + u_1^h + u_2^h$
end if

One final issue to be resolved in the filtering algorithm is how to split the residual vector r^h into r_1^h and r_2^h that contain the low and high frequency components respectively. Many choices are available. The filter used in [173] is the combination of a projection and a prolongation operators. The residual r^h is first projected to the coarse grid and then prolongated back to the fine grid again, i. e.

$$r_1^h = I_{2h}^h I_h^{2h} r^h.$$

The high frequency components in r are lost when r is projected to the coarse grid and then interpolated back again to the fine grid. The vector r_2^h can be calculated using $r_2^h = r^h - r_1^h$. Another alternative filter is given in [32].

As for the parallel implementations, both the fine grid correction and the coarse grid correction should be carried out simultaneously. Tuminaro [173] proposed an implementation for a two-dimensional $m \times m$ domain on a hypercube. The advantage of using hypercubes is that the architecture has 2^p nodes each of which is connected to p other nodes. Therefore, many communication steps involved in the filtering multigrid method are local in nature. At a particular grid G^h and a given cube of C^h, the implementation can be described as

Algorithm 2.7
Parallel Filtering Algorithm

1. Relax $A^h u^h = b^h$ on hypercube C^h

2. Calculate $r^h = b^h - A^h u^h$ on C^h

3. Split $r^h = r_1^h + r_2^h$ on C^h
   Calculate $r_1^{2h} = I_h^{2h} r_1^h$

4. Solve $A^h e_2^h = r_2^h$ on $C^h/2$

5. Solve $A^{2h} e_1^{2h} = r_1^{2h}$ on $C^h/4$

6. Update $u^h = u^h + I_{2h}^h e_1^{2h} + e_2^h$ on C^h

where $C^h, C^h/2$ and $C^h/4$ represent the original hypercube when the filtering multigrid method started on grid G^h, the sub-hypercube with half of the processors of C^h, and the sub-hypercube with a quarter of the number of processors of C^h, respectively. Communications are needed to move the fine grid correction problem to one-half of the original hypercube and to move the coarse grid correction problem to another quarter of the hypercube. If all grid points are properly mapped to processors, say by Gray code, then the communications are local. Also, since steps 4 and 5 are carried out simultaneously on different subcubes, there will be no contention of communication channels for the two correction problems. Since a subcube is still a hypercube, the division process can be defined recursively by replacing step 5 by "solve $A^{2h}e_1^{2h} = r_1^{2h}$ using filtering multigrid algorithm on $C^h/4$".

2.4.5   Concurrent Relaxations on Multiple Grids

In addition to splitting the original grid and filtering residual vector r to exploit parallelism, several researchers have proposed concurrent relaxations on multiple grids. These multiple grids can be several coarse grids at the same level [56, 67], or multiple grids at different levels [58]. Here we briefly discuss the parallel superconvergent multigrid proposed by Frederickson and McBryan [55, 56, 57, 125]. The basic idea of the parallel superconvergent multigrid is to use multiple coarse grids at a given grid level (a given spacing h) to generate a better correction for the solution at the fine grid than that generated by a single coarse grid in the traditional multigrid, so that the solution process converges fast. The computations for the multiple coarse grids do not require more time than what is needed for the single coarse grid, since the extra-coarse grid problems are solved on processors which would otherwise have been idle if only one coarse grid is used in the computation.

The process can be demonstrated using a one-dimensional grid [127]. Fig. 2.29(a) shows a one-dimensional grid which is to be coarsened in the multigrid solution process. In the standard multigrid method, the coarse grid is constructed by doubling the grid spacing h, so that all points with even numbers are left in the coarse grid, as shown in Fig. 2.29(b). The error equation is then solved on that coarse grid to produce a correction term for the approximate solution obtained at the fine grid. The other coarse grid containing all points with odd numbers as shown in Fig. 2.29(c) is not used in the standard multigrid method. Assume that in the multigrid solution process, the current fine grid contains just as many grid points as there are processors (this can happen either when there are many processors on a parallel computer, for example up to 65536 processors on a CM-2, or when the initial grid has been coarsened enough times so that there are not many points left in the current fine grid), then only about half of the processors will stay busy working on the next coarse grid shown in Fig. 2.29(b). The other half of the processors will be idle. The question then is whether we can utilize the idle processors to do some useful work on the other coarse grid shown in Fig. 2.29(c), so that the combined corrections from both coarse grids is better than that computed using either of the coarse grid separately.

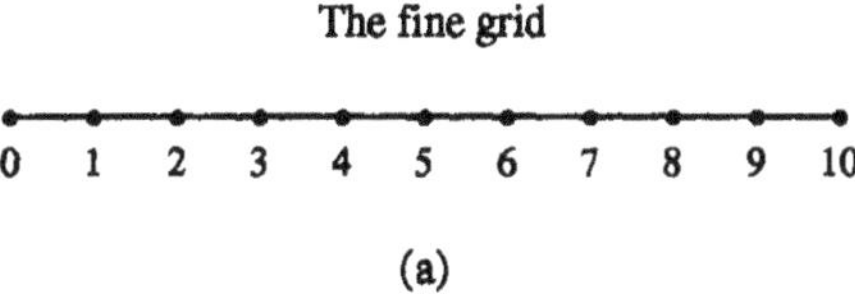

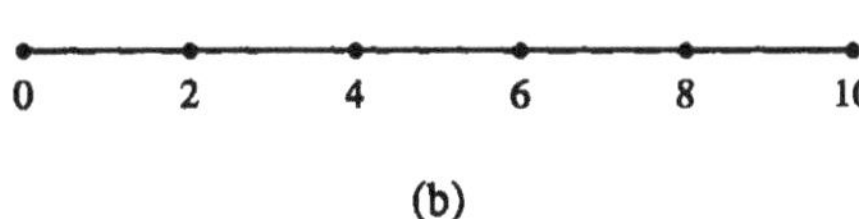

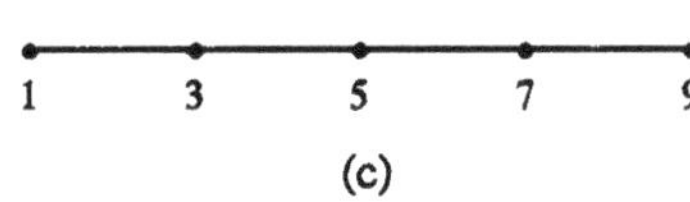

Figure 2.29: (a) The fine grid. (b) The coarse grid in the standard multigrid method. It contains all points with even numbers from the fine grid. (c) The extra coarse grid used in the superconvergent multigrid method. It contains all points with odd numbers.

This is possible since when the residual of the fine grid is projected to coarse grids, the even and odd numbered points receives, from the fine grid, different information which may complement each other. Therefore, a combination of the solutions from both coarse grids may represent the fine grid problem better than using either of the coarse grids alone.

In the case of higher dimensional problems, there are even more coarse grids at each given level that can be processed in parallel. For example, if the current fine grid is a two-dimensional 7×7 mesh as shown in Fig. 2.30(a), the coarse grid used

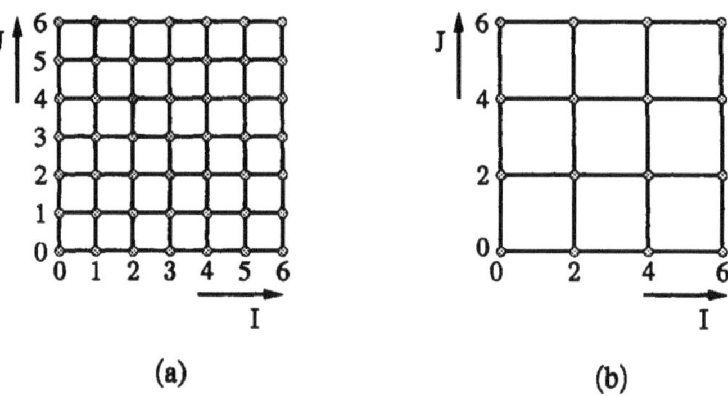

Figure 2.30: (a) A two-dimensional 7×7 mesh. (b) The coarse grid used in the standard multigrid method. All points have even i and j indices

in the standard multigrid method is shown in Fig. 2.30(b) which contains all grid points from the original fine grid with even i and j indices. This grid has only a third of all grid points in the original fine grid. The rest of the points can form three extra coarse grids shown in Fig. 2.31(a), (b) and (c) that can be used in the parallel superconvergent multigrid algorithm. In Fig. 2.31, all values of I and J indices are from the original fine grid.

In general, a fine grid in a d-dimensional space can be easily coarsened into 2^d coarse grids which can be used in the parallel superconvergent multigrid algorithm. Most of the basic operations in the standard multigrid algorithm, including smoothing and projection, still applies in the parallel superconvergent multigrid algorithm. The interpolation operation that brings the coarse grid corrections to the fine grid, however, needs to be chosen in such a way as to combine the correction terms from all

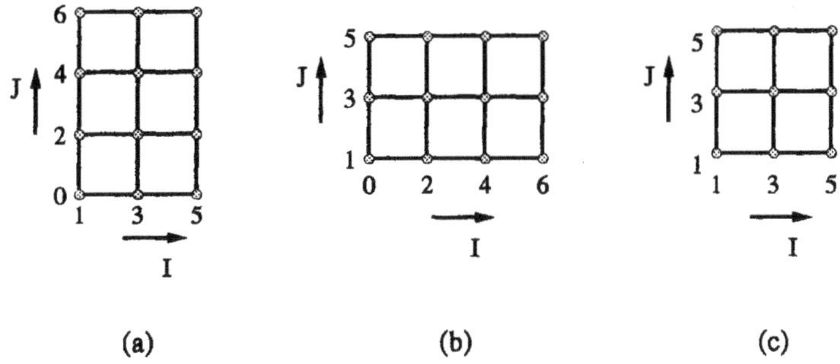

Figure 2.31: (a) A coarse grid with odd I index and even J index. (b) A coarse grid with even I index and odd J index. (c) A coarse grid with odd I and J indices.

coarse grids to optimize the convergence rate. We see from the previous analysis that the union of the coarse grids used in parallel superconvergent multigrid actually covers the whole fine grid, so the correction values (from different coarse grid solutions) are available at all grid points of the fine grid. Therefore, a trivial interpolation will be the identity matrix that does not change the correction values at any gird point. A better choice can be obtained using the Fourier mode analysis to optimize the convergence rate. The interpolation operators for many different cases are given in [56] and [127]. For example, for the Poisson's equation defined on a two-dimensional domain and discretized by the five-point central finite differencing scheme, the 9-point interpolation operator is given by

$$
Q_9 = \begin{bmatrix} q_{11} & q_1 & q_{11} \\ q_1 & q_0 & q_1 \\ q_{11} & q_1 & q_{11} \end{bmatrix}
$$

with $q_0 = 0.25$, $q_1 = 0.125$, $q_{11} = 0.0625$, and the 25-point interpolation operator is given by

$$
Q_{25} = \begin{bmatrix} q_{22} & q_{12} & q_2 & q_{12} & q_{22} \\ q_{12} & q_{11} & q_1 & q_{11} & q_{12} \\ q_2 & q_1 & q_0 & q_1 & q_2 \\ q_{12} & q_{11} & q_1 & q_{11} & q_{12} \\ q_{22} & q_{12} & q_2 & q_{12} & q_{22} \end{bmatrix}
$$

with $q_0 = 0.361017$, $q_1 = 0.11458$, $q_{11} = 0.0625$, $q_2 = -0.0309162$, $q_{12} = 0.00521024$, $q_{22} = 0.00316188$.

The action of the operator Q is interpreted in the same way as the finite difference stencils. Therefore, at point (i,j), the interpolated correction value e_{ij} is calculated by

$$e_{ij} = Q_9 v = q_0 v_{ij} + q_1(v_{i-1,j} + v_{i+1,j} + v_{i,j-1} + v_{i,j+1}) +$$
$$+ q_{11}(v_{i-1,j-1} + v_{i+1,j-1} + v_{i-1,j+1} + v_{i+1,j+1})$$

where $v_{i,j}$'s are the correction values calculated from different coarse girds. The calculation with Q_{25} is similar. For details of the Fourier mode analysis and the optimization of the convergence rate, readers are referred to [56] and [127].

The parallel superconvergent multigrid algorithm can be summarized as

---

Algorithm 2.8
Parallel Superconvergent Multigrid
Algorithm PSMG(h, u^h, b^h)

1. iterate ν_1 times on $A^h u^h = b^h$
2. compute residual $r^h = b^h - A^h u^h$
3. project residual to different coarse grids
4. do in parallel for all coarse grids G_k^{2h}
      solve for correction e_k^{2h}
      using PSMG($2h$, e_k^{2h}, r_k^{2h})
   end do
5. update u^h by $u^h = u^h + Q e_k^{2h}$
6. iterate ν_2 times on $A^h u^h = b^h$

As pointed out in [127], the parallel superconvergent multigrid algorithm can achieve the best results on massively parallel computers on which the number of processors is about the same as the number of grid points in the finest grid used in the computation. In [86], it was concluded that the parallel superconvergent multigrid method is close to four times more efficient than the standard multigrid method for many applications.

2.5   Parallel Conjugate Gradient Algorithm

In the previous sections, many algorithms have been discussed for solving the linear algebraic equation system

$$Ax = b \tag{2.55}$$

where A is a $n \times n$ matrix. A different approach is presented in this section in which the solution process is formulated as a minimization process for the case when A is symmetric and positive definite. The function to be minimized is

$$Q(x) = \frac{1}{2}x^T A x - b^T x. \tag{2.56}$$

The minimum of Eq. (2.56) is reached at the solution x^* of Eq. (2.55). The minimization process of (2.56) usually starts with an initial vector x^0, which is then updated repeatedly by

$$x^{k+1} = x^k + \alpha_k p_k, \qquad k = 0, 1, 2, \cdots, \tag{2.57}$$

to get a better solution. The vector p_k is a search direction along which $Q(x)$ is decreasing and α_k is the step size that determines how far the search should go in the direction p_k. Many different algorithms are available for solving this line search problem [66, 87, 143].

2.5.1   Conjugate Gradient Method

If the search vectors $p_0, p_1, \cdots, p_{n-1}$ in Eq. (2.57) are selected in such a way that

$$(p_i)^T A p_j = 0, \quad i \neq j, \tag{2.58}$$

then the method is called the conjugate gradient method which was first proposed by Hestenes and Stiefel [87]. The vectors p_i's are orthogonal with respect to the inner product defined by matrix A as $\langle x, y \rangle = x^T A y$.

An interesting property is that for a real $n \times n$ symmetric positive definite matrix A in Eq. (2.55), if the search directions p_i as in (2.58) are used and the α_k in $x^{k+1} = x^k + \alpha_k p^k$ is calculated to minimize the function (2.56), then the iterations defined by Eq. (2.57) converge to the true solution of (2.55) with any initial vector x_0 in no more than n steps. For proofs, see [143]. However, practically this property is not very attractive for the solution of large systems for two reasons: First, with a practical problem involving thousands of unknowns, one would expect the iterations to give a reasonable approximate solution in fewer than n iterations if the method is to be competitive. Second, the iteration will not converge to the exact solution even after n steps in the presence of round-off errors. Therefore the method has been used as an iterative method, rather than a direct method.

To determine α_k, substitute (2.57) into (2.56) and minimize

$$Q(\alpha_k) = \frac{1}{2}(x_k + \alpha_k p_k)^T A(x_k + \alpha_k p_k) - b^T(x_k + \alpha_k p_k) \qquad (2.59)$$

with respect to α_k, we obtain

$$\alpha_k = -\frac{p_k^T(Ax^k - b)}{(p^k)^T A p^k}. \qquad (2.60)$$

Since A is symmetric positive definite, the quantity $(p^k)^T A p^k$ is well defined and it is easy to verify that the minimum of $Q(\alpha_k)$ is reached at the value given by Eq. (2.60).

The algorithm for the conjugate gradient method can then be expressed as:

Algorithm 2.9
Conjugate Gradient Algorithm

select initial vector x_0
set initial search vector $p_0 = r_0 = Ax_0 - b$
compute $\gamma_0 = r_0^T r_0$
for $k = 0, 1 \cdots$

    1. $z_k = Ap_k$

    2. $\alpha_k = -\dfrac{\gamma_k}{p_k^T \cdot z_k}$

    3. $x_{k+1} = x_k + \alpha_k p_k$

    4. $r_{k+1} = r_k + \alpha_k z_k$

    5. if $\gamma_{k+1}^2 = r_{k+1}^T r_{k+1} \geq \varepsilon$ then
            continue
      else
            stop
      end if

    6. $\beta_k = \dfrac{\gamma_{k+1}^2}{\gamma_k^2}$

    7. $p_{k+1} = r_{k+1} + \beta_k p_k$

end for

Note in the above algorithm, step 4 calculates the new residual vector which is actually $r_{k+1} = r_k + \alpha_k z_k = Ax_k - b + \alpha_k Ap_k = A(x_k + \alpha_k p_k) - b = Ax_{k+1} - b$. Using $\alpha_k z_k$ rather than Ax_{k+1} can save one matrix vector multiplication. Step 6 and 7 calculate the new search direction. Readers are referred to [143] for the proof that the p_k's generated in this way are indeed orthogonal with respect to matrix A. In the conjugate gradient algorithm, it is clear that the major computational steps are the matrix-vector product Ap_k in step 1, inner products $p_k^T z_k$ and $r_{k+1}^T r_{k+1}$ in step 2 and 5, and vector updates for x_{k+1}, r_{k+1} and p_{k+1} in step 3, 4 and 7. Note that there are several other equivalent formulations for the conjugate gradient algorithm, for example that in [143].

2.5.2 Parallel Conjugate Gradient Algorithm

Since there are three computational components (matrix-vector product, inner product and vector update) in the conjugate gradient algorithm, we discuss separately parallelizations for each of these components. The first thing that needs to be considered is how the components in x_k (and p_k, r_k, z_k as well) are distributed among different processors.

1. Since each row of matrix A in Eq. (2.55) represents a discretized differential equation at a grid point in the discretized spatial domain, it is natural to decompose the whole spatial grid into subdomains which are assigned to different processors as shown in Fig. 2.4. This actually corresponds to distributing sets of rows of matrix A in Eq. (2.55) to different processors. Whether a set of rows assigned to a particular processor contains contiguous rows in matrix A depends on the shape of the spatial domain and the ordering of the grid points. For example, for the two-dimensional $m \times m$ domain shown ing Fig. 2.32, if the grid points are ordered in the I direction first, followed by the J direction, and each processor is assigned a strip of dimensions $\frac{m}{P} \times m$ were P is the number of processors, then processor P_i will be assigned rows in A numbered $(j-1)m + (i-1)\frac{m}{P} + k$, $k = 1, \cdots, \frac{m}{P}$, $j = 1, \cdots, m$. Here we have assumed that m is divisible by P. It can be seen that in this case not all rows of matrix A assigned to a processor are contiguous rows. If the grid points are ordered in the J direction first, followed by the I direction, then processor P_i will be assigned rows in matrix A numbered $(i-1)\frac{m^2}{p} + j$, $j = 1, \cdots, \frac{m^2}{p}$. Thus, a set of contiguous rows in matrix A is assigned to a processor.

Corresponding to the row assignments, all components in vector x_k (hence r_k, p_k and z_k as well) associated with the grid points of the particular subdomain should also be assigned to that particular processor. Additionally, to carry out the matrix product Ap_k, components of p_k at the grid points outside the particular subdomain also need to be available at the particular processor since the discretized equation at

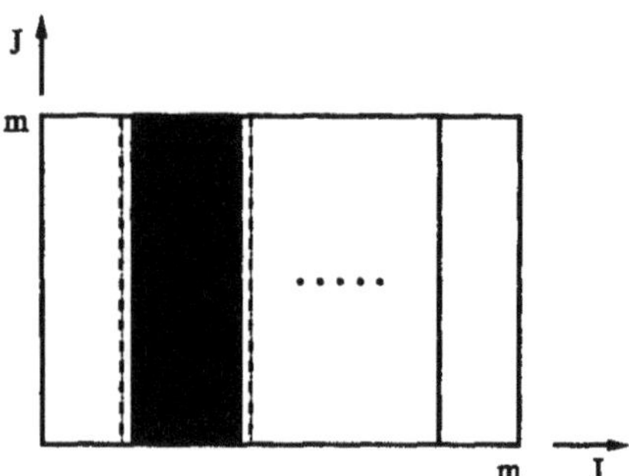

Figure 2.32: A processor is assigned a strip of the original domain. The dashed line indicates that values outside the subdomain are needed for the computation.

the boundary of the subdomain may involve unknowns outside the subdomain. This is accomplished by message exchange. The amount of message to be exchanged depends on the particular PDE and the stencil used in the discretization. For a second order elliptic PDE discretized by the central finite difference stencil as shown in Fig. 2.10(a), the components of p_k at one neighboring grid line on each side of the subdomain are needed for the computation of product Ap_k. Once all the components of p_k outside the particular subdomain are available on a particular processor, products of the rows of matrix A with the corresponding components in p_k can be carried out to compute the components of z_k associated with the grid points of that subdomain. Therefore, the product Ap_k can be parallelized very well once A and p_k are distributed to different processors.

2. For the two inner products $p_k^T \cdot z_k$ and $r_{k+1}^T \cdot r_{k+1}$, since each processor holds a section of vectors p_k, r_{k+1} and z_k, the partial sums of these inner products can be calculated in parallel by different processors. The partial sums need to be added together to get the final values of the inner products which should then be made available to all processors for vector updates. The computation and collection of partial sums for an inner product on 2^q processors can be carried out using the binary fan-in algorithm [143] as shown in Fig. 2.33. All processors calculate the partial sums $S_1, \cdots, S_8$ in parallel, then all partial sums are gathered to calculate the final value. For $P = 2^q$ processors, totally $q = \log_2 P$ steps are needed in the fan-in algorithm to get the final value of an inner product. As can be seen easily, for each

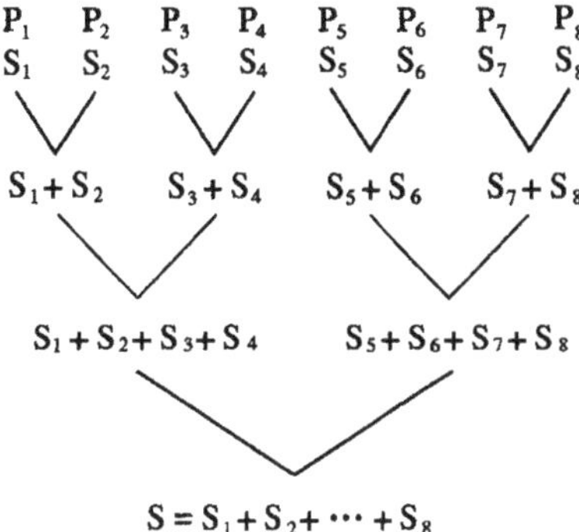

Figure 2.33: The fan-in algorithm for computing and collecting partial sums on 8 processors. P_i: processors, S_i: partial sums.

step down the tree, the degree of parallelism is reduced by half. For example, when all processors finish computing $S_1, \cdots, S_8$, the even numbered processors send their partial sums to odd numbered processors and then stay idle until the inner product is finally calculated. The rest of the computations are all done by odd numbered processors. Eventually the value of the inner product will be held by processor P_1 which then sends the value back to all processors.

On many of todays distributed memory parallel computers, processors are connected by bi-directional channels which make it possible for two processors to exchange messages simultaneously. A better implementation for an inner product utilizing this feature is to exchange, at each step of the fan-in algorithm, partial sums of two processors and then compute the next partial sum on both processors. For example, at the first step, processors P_1 and P_2 exchange partial sums S_1 and S_2, instead of just sending S_2 to P_1 from P_2, then both processors compute $S_1 + S_2$. In the end of the first step, P_1 and P_2 have the same value $S_1 + S_2$, P_3 and P_4 have the same value $S_3 + S_4$, similar results hold for other pairs of processors. For the next step, the pairs of processors are chosen as $P_1 \leftrightarrow P_3$, $P_2 \leftrightarrow P_4$, $P_5 \leftrightarrow P_7$ and $P_6 \leftrightarrow P_8$. So in the end of the second step, processor P_1, P_2, P_3, and P_4 will all have the partial sum $S_1 + S_2 + S_3 + S_4$, and the other processors will all have the partial sum $S_5 + S_6 + S_7 + S_8$. The process continues until the final sum is computed, but at each step, processors must be properly paired to exchange messages and compute next partial sums. When the computation is finished, the inner product will be available on all processors, thus saving the time for one processor to send the final results back to all processors as in the first implementation. This algorithm is particularly suitable for hypercube parallel computers.

3. Once the matrix-vector product and the two inner products are parallelized, the remaining component in a conjugate gradient iteration, i.e. the three vector updates in step 3, 4 and 7, can be done straight-forwardly in parallel, since the constants α_k and β_k should be available on all processors after the inner products have been calculated, and each processor should have a section of the vectors x_k, p_k, z_k, r_k and r_{k+1}.

The parallel Conjugate Gradient algorithm can now be given as:

Algorithm 2.10
Parallel Conjugate Gradient Algorithm

select initial guess x_0, $p_0 = r_0 = Ax_0 - b$, $\gamma_0 = r_0^T r_0$
for $k = 0, 1 \cdots$

    1. compute $z_k = Ap_k$ in parallel

    2. compute $p_k^T \cdot z_k$ by the fan-in algorithm, $\alpha_k = -\dfrac{\gamma_k}{p_k^T \cdot z_k}$

    3. update $x_{k+1} = x_k + \alpha_k p_k$ in parallel

    4. calculate $r_{k+1} = r_k + \alpha_k z_k$ in parallel

    5. calculate $\gamma_{k+1} = r_{k+1}^T \cdot r_{k+1}$ by the fan-in algorithm

    if $\gamma_{k+1} \leq \varepsilon$ then
        stop
    end if

    6. all processors calculate $\beta_k = \dfrac{\gamma_{k+1}}{\gamma_k}$

    7. calculate $p_{k+1} = r_{k+1} + \beta_k p_k$ in parallel

    8. exchange components of p_{k+1} between processors

end for

2.5.3   Preconditioned Conjugate Gradient Method (PCG)

As discussed in Section 2.5.1, the conjugate gradient method is used in practice as an iterative method for solving linear algebraic equation systems, although in theory it converges to the true solution in finite number of iterations with the exact arithmetic. It has been shown that the convergence rate of the Conjugate Gradient method is [123, 143]

$$\rho = \frac{\sqrt{\kappa} - 1}{\sqrt{\kappa} + 1} \tag{2.61}$$

where κ is the condition number of the coefficient matrix A

$$\kappa = cond(A) = \|A\| \cdot \|A^{-1}\|.$$

If the ℓ_2 matrix norm is used, the condition number κ will simply be the ratio of the largest and the smallest eigenvalues of A. It is clear from (2.61) that the closer the κ is to 1, the smaller the ρ is, and the faster the iterations converge.

In practical situations, the linear equation systems resulting from the discretization of PDEs usually do not possess the desired spectral distributions and condition numbers. Therefore, straightforward application of the CG method will not produce iterations with satisfactory convergence rate. This motivated the use of the Preconditioned Conjugate Gradient (PCG) method [12, 39, 86]. In the PCG method, a new equation system

$$\tilde{A}\tilde{x} = \tilde{b} \tag{2.62}$$

is solved to get a solution for the system

$$Ax = b, \tag{2.63}$$

where we have

$$\begin{aligned}
\tilde{A} &= MAM^T, \\
\tilde{x} &= M^{-T}x, \\
\tilde{b} &= Mb.
\end{aligned} \tag{2.64}$$

The matrix M should be chosen in such a way as to reduce the condition number of the coefficient matrix $\tilde{A}$, i.e. $\tilde{\kappa} = cond(\tilde{A}) < cond(A) = \kappa$, so that the CG iteration for the new system (2.62) converges faster. Eq. (2.62) is the preconditioned system. The CG method described before can be directly applied to (2.62). The solution obtained can then be transformed by (2.64) to get the solution of the original equation (2.63). However, in practical computations, it is desirable to avoid the explicit formation of $\tilde{A}$, $\tilde{x}$ and $\tilde{b}$ by using M and M^{-T}, since the operations alter the sparse structure of matrix A and can be costly to perform. By introducing a new vector

$$\tilde{r}_k = M^T M r_k$$

where $\tilde{r}_k$ is practically computed by solving

$$C\tilde{r}_k = r_k$$

with $C = (M^T M)^{-1}$, it can be shown that the CG algorithm applied to Eq. (2.62) can still be carried out in terms of the matrix and vectors in Eq. (2.63) as:

Algorithm 2.11

PCG Algorithm

select initial vector x_0, $r_0 = b - Ax_0$
solve $C\tilde{r}_0 = r_0$ and set $p_0 = \tilde{r}_0$, $\gamma_0 = \tilde{r}_0^T \cdot r_0$
for $k = 0, 1 \cdots$

1. $z_k = Ap_k$

2. $\alpha_k = -\dfrac{\gamma_k}{p_k^T \cdot z_k}$

3. $x_{k+1} = x_k + \alpha_k p_k$

4. $r_{k+1} = r_k + \alpha_k z_k$

5. solve $C\tilde{r}_{k+1} = r_{k+1}$

6. $\gamma_{k+1} = \tilde{r}_{k+1}^T \cdot r_{k+1}$

    if converged then
      stop
    end if

7. $\beta_k = \dfrac{\gamma_{k+1}}{\gamma_k}$

8. $p_{k+1} = \tilde{r}_{k+1} + \beta_k p_k$

end for

Note that in the above algorithm, if $C = I$, then the PCG algorithm reduces to the normal CG algorithm. If $C = A$, then for A symmetric positive definite, we can show that $\tilde{A} = I$, so the PCG iteration for Eq. (2.62) converges in just one trivial iteration. However the calculation of the preconditioning step $C\tilde{r}_0 = r_0$ will then be as difficult as that of solving the original equation (2.63). The central part in the PCG method is to construct a proper preconditioning matrix C such that the auxiliary system $C\tilde{r}_{k+1} = r_{k+1}$ is easy to solve and the preconditioned matrix (MAM^T) has a much smaller condition number than the original matrix A.

All steps, except the preconditioning step, in the PCG algorithm can be parallelized in the same way as discussed for the CG algorithm. The next several sections discuss parallelization of the preconditioning step which is often the bottle neck of the PCG method on parallel computers.

2.5.3.1 Truncated Series and Polynomial Preconditioning

The central part in the preconditioning step is to find a matrix C which is reasonably close to A and the auxiliary equation system

$$C\tilde{r}_{k+1} = r_{k+1},\qquad(2.65)$$

should be easy to solve. Since $\tilde{r}_{k+1}$ can also be written as

$$\tilde{r}_{k+1} = C^{-1}r_{k+1},\qquad(2.66)$$

the preconditioning step can be viewed as to calculate $\tilde{r}_{k+1}$ by the product of an approximation to A^{-1} with r_{k+1}. If the matrix A which is nonsingular is split as $A = P - Q$ where P is nonsingular and the largest eigenvalue in magnitude of $P^{-1}Q$ is less then 1, then the inverse of A can be expressed as [143]:

$$A^{-1} = (I + H + H^2 + H^3 + \cdots)P^{-1}\qquad(2.67)$$

where $H = P^{-1}Q$. A natural choice of an approximation to A^{-1} is to truncate the infinite series in (2.67) and set

$$C^{-1} = (I + H + H^2 + \cdots + H^m)P^{-1}.\qquad(2.68)$$

With this truncated series, Eq. (2.66) can be written as

$$\tilde{r}_{k+1} = (I + H + H^2 + \cdots + H^m)P^{-1}r_{k+1}.\qquad(2.69)$$

The actual computation in (2.69) is not to be carried out by explicitly forming the products of the matrix H. We note that Eq. (2.69) can be written equivalently as

$$\tilde{r}_{k+1} = (H(\cdots H(H + I) + \cdots + I) + I)P^{-1}r_{k+1}$$

where there are m nested parentheses, so that

$$\tilde{r}_{k+1} = (H(\cdots H(HP^{-1}r_{k+1} + P^{-1}r_{k+1}) + \cdots + P^{-1}r_{k+1}) + P^{-1}r_{k+1}).\qquad(2.70)$$

Let

$$r_{k+1}^0 = 0,\quad r_{k+1}^1 = P^{-1}r_{k+1},$$

then $\tilde{r}_{k+1}$ can be calculated recursively as

$$r_{k+1}^{i+1} = Hr_{k+1}^{i} + r_{k+1}^{1}, \qquad i = 0, 1, \cdots, m-1. \tag{2.71}$$

Recall that $H = P^{-1}Q$, we then have

$$Pr_{k+1}^{i+1} = Qr_{k+1}^{i} + r_{k+1}, \qquad i = 0, 1, \cdots, m-1, \tag{2.72}$$

with $\tilde{r}_{k+1} = r_{k+1}^{m}$. Eq. (2.72) shows that preconditioning by a truncated matrix series of order m represented by (2.68) is actually equivalent to performing m iterations given by Eq. (2.72) with the residual vector r_{k+1} as the right hand side term. Eq. (2.72) reminds us of the familiar relaxation algorithms discussed in section 2.2. If A is split as $A = D - L - U$ with D the diagonal part, $-L$ and $-U$ the strictly lower and upper triangular parts of A respectively, then setting $P = D$ and $Q = L + U$ corresponds to the Jacobi iterations in Eq. (2.72) and setting $P = D - L$ and $Q = U$ corresponds to the Gauss-Seidel iterations in Eq. (2.72). These two strategies are called the Jacobi and Gauss-Seidel preconditioning respectively. The SOR preconditioning procedure can be defined based on the Gauss-Seidel preconditioning. The problem with the G-S or SOR preconditioning is that the preconditioning matrix C, or C^{-1} given by (2.72), is in general not symmetric. To overcome this difficulty, the symmetric SOR method (SSOR) can be used for the iterations in (2.72). It has been shown [143] that with A symmetric positive definite and $0 < \omega < 2$, the preconditioning matrix C and C^{-1} in (2.72) with SSOR iterations is also symmetric positive definite.

The advantage of using truncated series preconditioning is that the procedure is finally reduced to performing Jacobi or SSOR iterations for the residual vector r_{k+1}. As has been discussed in section 2.2.3, the Jacobi iterations can be parallelized straightforwardly and the SSOR iterations can be parallelized using red-black ordering or multicoloring schemes. Therefore, the truncated series preconditioning procedure is highly parallelizable.

For the implementation, the parallel PCG method will be very similar to the parallel CG method discussed before. All steps in the PCG method can be parallelized in the same way as has been used for the CG method, except for step 5 — the preconditioning step where the statement

$$\text{5. solve } C\tilde{r}_{k+1} = r_{k+1}$$

should be changed to

$$\text{5. Carry out } m \text{ Jacobi (or SSOR) iterations on } A\tilde{r}_{k+1} = r_{k+1}.$$

In the end of the parallel PCG algorithm, the components in p_{k+1} associated with the grid points on the boundary of a subdomain should be exchanged with the neighboring subdomains holding adjacent subdomains. Additionally, to carry out the relaxations for the preconditioning step, the components in the vector r_{k+1} associated with the grid points on the boundary of a subdomain should also be exchanged with the neighboring processors holding adjacent subdomains. The parallel truncated series PCG algorithm can now be given as:

Algorithm 2.12
Parallel Truncated Series PCG

select x_0 and compute $r_0 = Ax_0 - b$
relax m times on $A\tilde{r}_0 = r_0$
set $p_0 = \tilde{r}_0$, and $\gamma_0 = \tilde{r}_0^T \cdot r_0$
for $k = 0, 1 \cdots$

1. compute $z_k = Ap_k$ in parallel

2. compute $p_k^T \cdot z_k$ using the fan-in algorithm, $\alpha_k = -\dfrac{\gamma_k}{p_k^T \cdot z_k}$

3. update $x_{k+1} = x_k + \alpha_k p_k$ in parallel

4. compute $r_{k+1} = r_k + \alpha_k z_k$ in parallel

5. exchange components of r_{k+1} on the boundary of subdomains

6. relax m times on $A\tilde{r}_{k+1} = r_{k+1}$ in parallel

7. computer $\gamma_{k+1} = \tilde{r}_{k+1}^T \cdot r_{k+1}$ using the fan-in algorithm

   if converged then
      stop
   end if

8. set $\beta_k = \dfrac{\gamma_{k+1}}{\gamma_k}$

9. compute $p_{k+1} = \tilde{r}_{k+1} + \beta_k p_k$ in parallel

10. exchange components in p_{k+1} on the boundary of subdomains.

end for

A potential problem with the truncated series preconditioning is that the relaxation process used in solving $A\tilde{r}_{k+1} = r_{k+1}$ does not converge for all matrices. It is guaranteed to converge for A symmetric positive definite. However, there are many cases (for example in diffusion-convection problems) where the discretized PDEs do not have a symmetric positive definite matrix A. Some research work has been reported (see [104], for example) recently to demonstrate the use of Jacobi and SSOR preconditioning in solving non-symmetric linear equations arising from discretized PDEs in meteorology.

2.5.3.2　Incomplete Factorization Preconditioning

The conjugate gradient method combined with the incomplete factorization as the preconditioning step is often called ICCG method, since most of the time the approximate factorization used is the Incomplete Cholesky factorization for symmetric positive definite matrices.

As discussed before, the preconditioning step is to find a proper matrix C close to A and to solve $C\tilde{r} = r$. One way to find such a matrix C is to use the incomplete factorization of A (The exact factorization is usually expensive to compute, otherwise we could have solved $Ax = b$ directly by a factorization method). For example, if A is symmetric positive definite, we can factorize A as

$$A \approx \tilde{A} = LL^T = C \tag{2.73}$$

where L is a lower triangular matrix and maintains a similar sparse structure like A. Note that the exact factorization usually produces a dense L. A straightforward incomplete Cholesky factorization scheme can be given as:

Algorithm 2.13
Incomplete Cholesky Factorization

$l_{11} = \sqrt{a_{11}}$
for $i = 2$ **to** n
　　for $j = 1$ **to** $i - 1$
　　　　if $(a_{ij} = 0)$ **then**
　　　　　　$l_{ij} = 0$
　　　　else
$$l_{ij} = \left(a_{ij} - \sum_{k=1}^{j-1} l_{ik} l_{jk} \right) / l_{jj}$$
　　　　end if
　　end for
$$l_{ii} = \sqrt{a_{ii} - \sum_{k=1}^{i-1} l_{ik}^2}$$
end for

The "**if**" statement in the above algorithm ensures that no fill-in will be produced in the factorized matrix L which will have the same sparse structure as matrix A. A significant amount of computation is also saved since only those elements l_{ij} in L that correspond to nonzero a_{ij}'s need to be computed. This no fill-in incomplete Cholesky factorization is also called IC(0) factorization. There are many other approximate factorization schemes that allow different levels of fill-in. These methods are usually referred to as IC(n) methods where n indicates the level of fill-in.

For more general matrix A, the approximate factorization will be in the form of

$$A \approx \tilde{A} = LU = C. \tag{2.74}$$

The schemes with no fill-in or different levels of fill-in are denoted by ILU(0) and ILU(n) respectively. More detailed discussions about different pre-conditioners using approximate factorizations can be found in [66, 99, 143, 154, 174, 195].

Once the approximate factorization is calculated, it will be used throughout the whole iteration process in the PCG method. In each PCG iteration, a forward and a backward substitutions are needed to solve Eq. (2.65). Therefore, it is important to parallelize the substitution processes for better algorithm performance.

Since the approximate factorization is sparse, the general parallel triangular system solver, for example that given in [79, 121], will not be efficient. There are many different ways to parallelize the solution process of $A\tilde{r} = r$ once the approximate factorization of A is available. We discuss here two ways for the algebraic equations derived from the discretization of the Poisson's equation on a two-dimensional $n \times n$ square domain. The standard central finite difference scheme is used to discretize the PDE and the no fill-in approximate factorization is used as the preconditioner. The original matrix A in the equation system

$$Ax = b$$

is penta-diagonal with the structure

$$\begin{bmatrix} A_1 & B_2^T & & & \\ B_1 & A_2 & B_3^T & & \\ & B_2 & A_3 & \ddots & \\ & & \ddots & \ddots & B_n^T \\ & & & B_{n-1} & A_n \end{bmatrix} \tag{2.75}$$

where A_i's are a tri-diagonal matrices and B_i's are diagonal matrices. If the no fill-in approximate factorization is used to factorize A as

$$A \approx \tilde{A} = LL^T,$$

then the preconditioning step is composed of solving

$$Ly = r \tag{2.76}$$

and

$$L^T \tilde{r} = y \tag{2.77}$$

where the matrix L has the structure shown in Fig. 2.34. There are totally $N = n \times n$

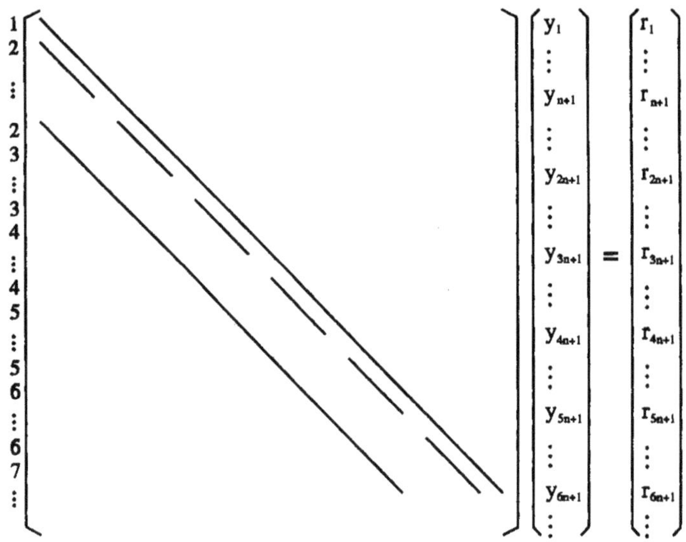

Figure 2.34: Forward substitution of a lower triangular system in the ICCG(0) method. The numbers on the left side of the matrix indicates the step at which the corresponding equation can be solved.

equations in the system. For the forward substitution, once the first equation is solved, the value of x_1 can be used to solve the second and the $n+1th$ equation for x_2 and x_{n+1}. Then the third, the $n+2th$ and the $2n+1th$ equations can be solved at the third step. It is clear that as the process progresses, an additional equation can be solved in parallel until the nth step at which n equations are solved in parallel. After that, the number of equations being solved in parallel will decrease since the equations in the first block have all been solved.

For parallel implementation, we can assign different blocks of unknowns to different processors and solve equations in different blocks in parallel. With n processors, the system can be solved in $2n$ steps, as compared to n^2 steps with one processor in the sequential case. We see that the speedup is less than $n/2$. This is because the load is not well balanced during the substitution process.

The backward substitution can be carried out in the same way except the process starts from the last equation and then proceeds bottom-up.

The extension of this approach to three-dimensional domains is straightforward, see [143] for more discussion and analysis about complicated equations which generate matrix A in (2.75) with penta-diagonal A_i and tri-diagonal B_i.

The other approach is to use red-black ordering (or multi-coloring for more complicated equations or difference schemes) to parallelize the forward and backward substitutions. This comes from the observation that the data dependence of the forward substitution is exactly the same as that in the Gauss-Seidel iterations. Therefore, the red-black ordering discussed in section 2.2.3 can be used to discretize the substitution process.

By using red-black ordering , the matrix in (2.75) will look like

$$\begin{bmatrix} D_r & B_{br}^T \\ B_{br} & D_b \end{bmatrix} \tag{2.78}$$

where D_r and D_b are diagonal matrices that correspond to the red and black unknowns, and B_{br} is a sparse matrix. If the level-0 incomplete factorization is used to factorize the matrix in (2.78) as

$$\begin{bmatrix} D_r & B_{br}^T \\ B_{br} & D_b \end{bmatrix} = \begin{bmatrix} D_1 & \\ C & D_2 \end{bmatrix} \begin{bmatrix} D_1 & C^T \\ & D_2 \end{bmatrix} \tag{2.79}$$

where we have the relations

$$D_1 \cdot D_1 = D_r, \quad C = B_{br} \cdot D_1^{-1}, \quad D_2 \cdot D_2 = D_b - C \cdot C^T.$$

The sub-matrices D_1 and C can be calculated easily since both D_1 and D_r are diagonal matrices with positive entries. In the computation for D_2 which is supposed to be a diagonal matrix too, only the part that generates the diagonal elements in D_2 needs to be carried out in the IC(0) factorization. Note the first two steps for computing D_1 and C are valid even in the case of exact factorization.

After the factorization shown in (2.79) has completed, the preconditioning step is then to solve

$$\begin{bmatrix} D_1 & \\ C & D_2 \end{bmatrix} \begin{bmatrix} y_r \\ y_b \end{bmatrix} = \begin{bmatrix} r_r \\ r_b \end{bmatrix} \tag{2.80}$$

and

$$\begin{bmatrix} D_1 & C^T \\ & D_2 \end{bmatrix} \begin{bmatrix} \tilde{r}_r \\ \tilde{r}_b \end{bmatrix} = \begin{bmatrix} y_r \\ y_b \end{bmatrix}. \tag{2.81}$$

It is obvious when solving Eq. (2.80) that y_r can be computed in parallel since D_1 is diagonal. After y_r is computed, y_b can also be computed in parallel by solving

$$D_2 y_b = r_b - C y_r$$

since matrix D_2 is also diagonal.

The solution process for Eq. (2.81) is the same as that for Eq. (2.80) except that the process goes backwards by first solving $\tilde{r}_b$ in parallel and then $\tilde{r}_r$.

For parallel implementation on distributed memory parallel computers, the original space grid on which the PDE is discretized should be decomposed into subgrids as discussed in section 2.5.2. The components in the corresponding solution, residual and search direction vectors associated with a subgrid should be assigned to the processor holding the subgrid. The matrix-vector products, inner products and vector updates should all be carried out as discussed in section 2.5.2. As for the preconditioning step, each subgrid contains approximately equal numbers of red and black points, so the substitutions for Eqs. (2.80) and (2.81) are first carried out on the points of one color, followed by the grid points of the other color. Note that after each substitution for a particular color, the updated values on the boundaries of subgrids should be exchanged with the neighboring processors holding adjacent subgrids, since the updated values are needed in the calculations for the points of the other color.

Similar to the situation discussed in section 2.2.3.2, multi-coloring can be used to parallelize the forward and backward substitutions of Eqs. (2.80) and (2.81) for more complicated equations discretized by high order difference schemes. If m colors are used, then each of the substitutions in Eqs. (2.80) and (2.81) involves m steps for different colors. The computations for the grid points of the same color can be done in parallel.

2.5.3.3   Domain Decomposition Preconditioning

The term "Domain Decomposition" has been used to represent the method for solving PDEs defined on a spatial domain by decomposing the original domain into subdomains. The sub-problems defined on the subdomains are solved separately and the final solution is obtained by combining the solutions from subdomains. The solution process can be iterative in nature. Although the idea of domain decomposition was proposed in the last century by Schwartz (see [41]), it is used today in a much broader context. One of the major advantages of domain decomposition method is the natural parallelism in solving the sub-problems defined on the subdomains. In our previous discussions, we have seen that the spatial grid on which the original PDE is defined is always decomposed into subgrids (can be considered as subdomains) which are then assigned to different processors for parallel processing.

In this section, we discuss the use of domain decomposition as a preconditioning step. For simplicity of the discussion, we assume that an elliptic PDE is discretized on a rectangular two-dimensional domain which is decomposed into two subdomains as shown in Fig. 2.35. If we number the grid points inside Ω_1 first, followed by the

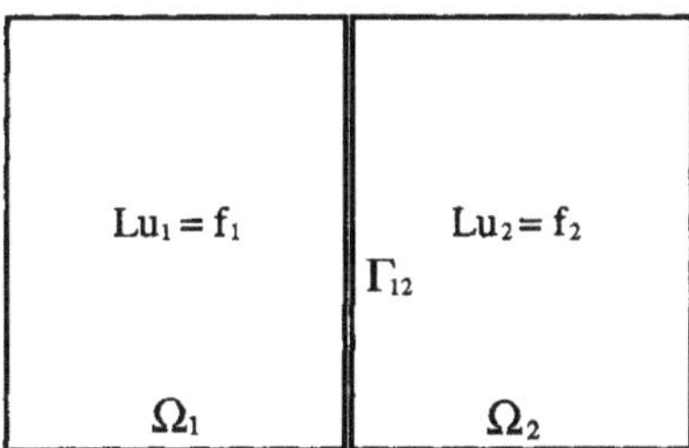

Figure 2.35: The original domain Ω is decomposed into subdomains Ω_1 and Ω_2. Correspondingly, the original PDE $Lu = f$ is represented as two subproblems $Lu_1 = f_1$ and $Lu_2 = f_2$ in two subdomains respectively. Γ_{12} is the boundary between the two subdomains

points inside Ω_2, and then finally number the grid points on the boundary Γ_{12} which separates Ω_1 and Ω_2, the discretized equations will be in the form of

$$Au = \begin{bmatrix} A_1 & & C_1^T \\ & A_2 & C_2^T \\ C_1 & C_2 & B_{12} \end{bmatrix} \begin{bmatrix} u_1 \\ u_2 \\ u_{12} \end{bmatrix} = \begin{bmatrix} f_1 \\ f_2 \\ f_{12} \end{bmatrix} = f \tag{2.82}$$

where u_1, u_2 and u_{12} represent the unknowns to be solved on Ω_1, Ω_2 and Γ_{12} respectively, A_1, A_2 and B_{12} are symmetric positive definite, and typically the order of B_{12} is much smaller than that of A_1 and A_2.

If Eq. (2.82) is to be solved by the Preconditioned Conjugate Gradient method, we need to find a preconditioning matrix C which is reasonably close to A and the equation system

$$C\tilde{r} = r$$

is easy to solve in the PCG iterations. Since the matrix in Eq. (2.82) is a block matrix, it is natural to consider a block approximate factorization of matrix A in the form of

$$C = L \begin{bmatrix} D_1^{-1} & & \\ & D_2^{-1} & \\ & & S^{-1} \end{bmatrix} L^T \tag{2.83}$$

with

$$L = \begin{bmatrix} D_1 & & \\ & D_2 & \\ C_1 & C_2 & S \end{bmatrix}. \tag{2.84}$$

It is easy to verify that

$$C = \begin{bmatrix} D_1 & & C_1^T \\ & D_2 & C_2^T \\ C_1 & C_2 & B \end{bmatrix}, \tag{2.85}$$

where $B = S + C_1 D_1^{-1} C_1^T + C_2 D_2^{-1} C_2^T$. For matrix C to be a good approximation to A in Eq. (2.82), we expect D_1 and D_2 to be close to A_1 and A_2, and B to be close to B_{12} which result in

$$S \approx B_{12} - C_1 D_1^{-1} C_1^T - C_2 D_2^{-1} C_2^T \tag{2.86}$$

Once the factorization shown in Eq. (2.83) is available, the preconditioning step can be carried out as

$$\begin{bmatrix} D_1 & & \\ & D_2 & \\ C_1 & C_2 & S \end{bmatrix} \begin{bmatrix} x_1 \\ x_2 \\ x_3 \end{bmatrix} = \begin{bmatrix} r_1 \\ r_2 \\ r_3 \end{bmatrix},$$

$$\begin{bmatrix} y_1 \\ y_2 \\ y_3 \end{bmatrix} = \begin{bmatrix} D_1 & & \\ & D_2 & \\ & & S \end{bmatrix} \begin{bmatrix} x_1 \\ x_2 \\ x_3 \end{bmatrix}, \tag{2.87}$$

$$\begin{bmatrix} D_1^T & & C_1^T \\ & D_2^T & C_2^T \\ & & S^T \end{bmatrix} \begin{bmatrix} \tilde{r}_1 \\ \tilde{r}_2 \\ \tilde{r}_3 \end{bmatrix} = \begin{bmatrix} y_1 \\ y_2 \\ y_3 \end{bmatrix}.$$

The first two steps $D_1 x_1 = r_1$ and $D_2 x_2 = r_2$ correspond to solving two subproblems in the two subdomains which can be carried out in parallel. The step $S x_3 = r_3 - C_1 x_1 - C_2 x_2$ has to wait until the first two steps finish. The computation for y can be done in parallel since it is a matrix-vector product. In the last stage, the equation $S^T \tilde{r}_3 = y_3$ must first be solved, then the last two blocks of equations $D_2^T \tilde{r}_2 = y_2 - C_2^T \tilde{r}_3$ and $D_1^T \tilde{r}_1 = y_1 - C_1^T \tilde{r}_3$ can be solved in parallel.

If the subdomains are rectangular and the PDE is not very complicated, then it is possible that the subdomain problems can be solved efficiently. In this case , D_1 and D_2 can be chosen as the submatrices A_1 and A_2 in the original matrix A of Eq. (2.82). The major part in the preconditioning step is then to solve the subproblems defined in the subdomains. The two steps for the boundary problem

$$Sx_3 = r_3 - C_1 x_1 - C_2 x_2 \tag{2.88}$$

and

$$S\tilde{r}_8 = y_3 \tag{2.89}$$

are in general more difficult to handle. The first problem is that it can not be processed simultaneously with the other subproblems defined in subdomains. Therefore, parallelism can only be exploited at a much fine level within the solution process for Eqs. (2.88) and (2.89). Fortunately the size of Eqs. (2.88) and (2.89) is much smaller than the subdomain problems , typically by an order of magnitude. The second problem is that the formation of matrix S as shown in Eq. (2.86) involves the matrix inversions D_1^{-1} and D_2^{-1} which we do not want to compute explicitly. There are two ways to get around this problem. One way is to use approximations rather than the exact inverses D_1^{-1} and D_2^{-1}. If the matrices D_1 and D_2 are symmetric positive definite, then D_1^{-1} and D_2^{-1} can be approximated by a diagonal, tri-diagonal, or banded matrices [40, 66] in the amount of computations proportional to the order of the matrices D_1 and D_2. This type of approximation not only reduces the amount of numerical computation, but also maintains the sparse structure in matrix S.

The other way to avoid the explicit formation of D_1^{-1} and D_2^{-1} is to use iterative methods to solve Eqs. (2.88) and (2.89). With iterative algorithms, we do not need to form the matrix S explicitly. What is needed is the product Sv with some vector v as shown in Step 1 of the conjugate gradient algorithm. To form

$$\begin{aligned} Sv &= (B_{12} - C_1 D_1^{-1} C_1^T - C_2 D_2^{-1} C_2^T)v \\ &= B_{12}v - C_1 D_1^{-1} C_1^T v - C_2 D_2^{-1} C_2^T v, \end{aligned}$$

we note that $D_1^{-1} C_1^T v$ and $D_2^{-1} C_2^T v$ can be formed by solving

$$D_1 z_1 = C_1^T v \tag{2.90}$$

and

$$D_2 z_2 = C_2^T v, \tag{2.91}$$

so that

$$Sv = B_{12}v - C_1 z_1 - C_2 z_2.$$

Here we have assumed that the subdomain problems (2.90) and (2.91) are easy to solve.

For more general situations where m subdomains are involved, by ordering the grid points inside each subdomain first and then the points on the boundaries between subdomains, we will have a matrix A with the structure similar to that in Eq. (2.82):

$$L = \begin{bmatrix} A_1 & & & & C_1 \\ & A_2 & & & C_2 \\ & & \ddots & & \vdots \\ & & & A_m & C_m \\ C_1 & C_2 & \cdots & C_m & B \end{bmatrix}. \tag{2.92}$$

The preconditioning matrix C can be written as

$$C = L \begin{bmatrix} D_1^{-1} & & & \\ & \ddots & & \\ & & D_m^{-1} & \\ & & & S^{-1} \end{bmatrix} L^T$$

with

$$L = \begin{bmatrix} D_1 & & & \\ & \ddots & & \\ & & D_m & \\ C_1 & \cdots & C_m & S \end{bmatrix}.$$

If each subproblem in a subdomain is easy to solve, then we can set $D_i = A_i$ and S will be

$$S = B - \sum_{i=1}^{m} C_i A_i^{-1} C_i^T \tag{2.93}$$

where A_i^{-1} can either be approximated by a banded matrix or replaced by its action on a vector v if the equation

$$Sx = r \tag{2.94}$$

is solved by iterative methods.

For parallel implementations, since all subproblems can be processed in parallel, each processor can be assigned a subdomain to carry out the forward and backward substitutions. All terms in the summation of Eq. (2.93) can also be computed in parallel by different processors holding different subdomains. As for Eq. (2.94) , depending on its size , it can either be solved on a single processor or distributed to multiple processors for parallel processing. The former is easier for implementation, but could be a potential bottle neck for parallel computing. The latter is more complicated in implementation and if the equation system is not large enough, the gain from parallel processing may not be enough to offset the loss due to the parallel overhead and additional communication cost.

Another simple and perfectly parallelizable domain decomposition preconditioning can be obtained by treating the boundary unknowns explicitly. Therefore, when forming the auxiliary preconditioning equation

$$C\tilde{r} = r. \tag{2.95}$$

the matrix C can be taken as the diagonal part of the matrix A in (2.92)

$$C = \begin{bmatrix} A_1 & & & \\ & A_2 & & \\ & & \ddots & \\ & & & A_m \end{bmatrix}, \tag{2.96}$$

and the components in $\tilde{r}$ corresponding to the submatrix B in (2.92) are held at their values from the previous iteration, or from r_0 for the first iteration. All subproblems $A_i\tilde{r} = r_i$ can then be solved in parallel by different processors. This method can also be viewed as a block Jacobi preconditioning [11]. While the method is straightforward and perfectly parallelizable, it usually does not improve the convergence rate significantly, even for the simple Poisson's equation [143].

Although the Conjugate Gradient algorithm has mainly been applied to algebraic equation systems with symmetric positive definite matrices, many researchers have also extended the method, both the CG method itself and the preconditioning step, to many non-symmetric or indefinite problems. Oppe and Kincaid [141] applied pre-conditioned Conjugate Gradient method, with both block Jacobi, SSOR, polynomial and incomplete Cholesky factorization preconditioners, to nonsymmetric equations arising from oil reservoir simulations; Wathen and Silvester [185] used Conjugate Gradient Method with simple diagonal preconditioning to solve Stokes problem describing slow viscous incompressible flow problem with symmetric indefinite equation systems. There are also different Conjugate Gradient-like methods, called Krylov space methods, for nonsymmetric methods, for example the Generalized Conjugate Gradient Method which was used by Evans and Li [48] to solve hyperbolic difference equations, and the GMRES method [155] with was used by Navarra [134] and Kadiŏglu and Mudrick [104] to solve nonsymmetric problems arising in meteorology, by Shakib [161] to solve compressible Euler and Navier-Stokes equations using finite element method, and by Tan and Bath [169] for solving structure and fluid flow problems with incomplete LU factorization preconditioning. Recent extensions of the GMRES method can be found in [132].

More detailed discussions on parallel implementations of the Conjugate Gradient type algorithms with different preconditioners on different architectures can be found in [128] in which parallel domain decomposition preconditioning was used on Cray and ETA parallel computers; in [52] in which the overlapped partitioned incomplete LU factorization preconditioner is used with the CG type algorithms for solving equations with general sparsity patterns on IBM 3090/600E parallel computers; in [37, 38] where the s-step conjugate gradient method with incomplete Cholesky factorization and polynomial preconditioning was implemented on the Alliant FX/8 parallel computer; in [140] where a parallel block conjugate gradient algorithm was analyzed and implemented on torus and hypercube parallel computers; in [47] where Elman and Agrón analyzed the incomplete factorization preconditioning with multicoloring on loosely coupled parallel computers and concluded that the use of multicolor ordering results in slower convergence rate than the natural ordering, but the gain from the parallelism of multicoloring is more than offsetting the loss due to the slower convergence rate, so that the overall performance on parallel computers is better; in [124] where Mansfield used a combination of damped Jacobi precondition and deflation to improve the convergence rate of the CG method for solving the linear system associated with the elastic bending of a cantilever beam on iPSC/2 hypercubes; and in [163] where Sobh discussed the CG algorithm with the point Jacobi preconditioning for solving the Navier equations in the elasticity theory on CM-2 parallel computers.

Chapter

3

Implementations

Many parallel algorithms have been discussed in Chapter 2. This chapter will address the implementation issues. In general, there are three main steps involved in solving a PDE related problem on parallel computers: Developing a parallel algorithm for the given problem, implementing the algorithm on a specific parallel computer, and fine-tuning the program for better efficiency. The second step gets a program running on a parallel computer, and the third step makes the program run faster. This process is usually architecture dependent. The discussions here are based on the Intel iPSC/860 and the KSR-1 parallel computers. While the technical details may apply only to the specific architectures discussed here, the basic concepts should be applicable to a much wider class of architectures.

3.1   Intel iPSC/860 Hypercubes

As discussed in Chapter 1, the hypercube architecture is a very special connection topology in which all processors are highly connected. The number of processors on a hypercube is always a power of 2. On a d-dimension hypercube with 2^d processors, each processor is directly connected to d other processors which are called the nearest neighbors. The processors on a hypercube are also called nodes. While it is easy to see that a zero-, one-, two-, or three-dimensional hypercube corresponds to the geometry of a point, a line, a square, or a cube respectively as shown in Fig. 3.1, it is hard to visualize higher dimensional hypercubes. One can think of the construction of a d-dimensional hypercube as connecting the corresponding nodes of two $(d-1)$-dimensional hypercubes. Fig. 3.2 shows that a four-dimensional hypercube is the combination of two three-dimensional hypercubes.

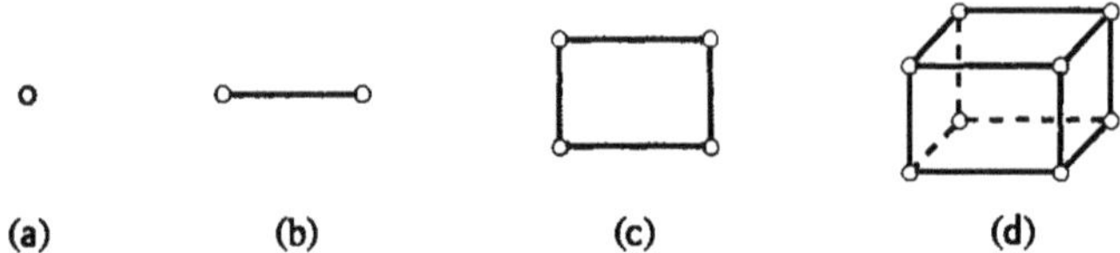

Figure 3.1: (a) a zero-dimensional hypercube, (b) a one-dimensional hypercube, (c) a two-dimensional hypercube, (d) a three-dimensional hypercube.

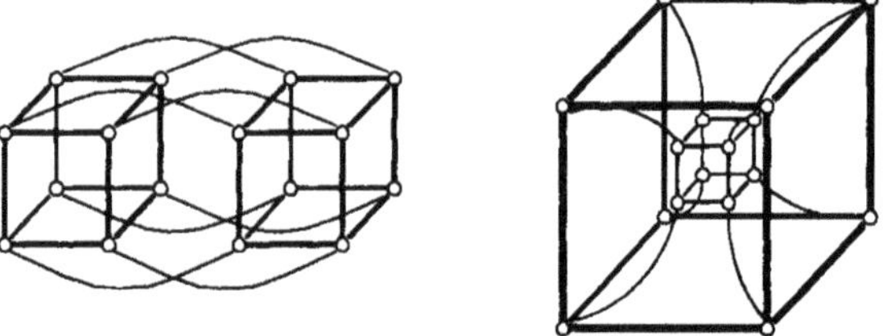

Figure 3.2: A four-dimensional hypercube is constructed by connecting all nodes of a three-dimensional hypercube with the corresponding nodes of another three-dimensional hypercube.

One of the most widely used hypercube parallel computers is the Intel iPSC/860 hypercube (1990). It is the third generation of hypercubes produced by Intel after the iPSC/1 and iPSC/2 hypercubes. The architecture can support up to 128 computational nodes (a 7-dimensional hypercube).

Each computational node on an iPSC/860 consists of an Intel i860 processor plus memory and communication components. At a clock rate of 40 MHz, each i860 processor has a peak execution rate of 32 MIPs (millions of instructions per second) integer performance, 80 MFlops (millions of floating point operations per second) 32-bit floating point performance which is often referred to as the single precision performance, and 60 MFlops 64-bit floating point performance which is often referred to as the double precision performance. Therefore, the aggregate peak performance of a 128-node iPSC/860 Intel hypercube is about 10.2 GFlops (billions of floating point operations per second) for single precision and 7.7 GFlops for double precision. These peak execution rates, however, are very difficult to achieve or sustain when solving practical problems. It is not unusual in practice to observe a snstained rate which is only 10 to 20 percent of the peak performance [78]. The performance degradation is mainly due to the cache miss, non-ideal instruction mix, improper data alignment, and non-vectorized computations.

The memory hierarchy of iPSC/860 consists of local cache and local memory. Each node has 8 Kbytes of fast cache and up to 64 Mbytes (could be 8, 16, 32Mbytes) of local memory. Therefore, the maximum configuration of an iPSC/860 hypercube can have 8 Gbytes of aggregate distributed memory.

In addition to the computational nodes, an iPSC/860 hypercube can also have several I/O nodes. Each I/O node consists of an Intel 80386 processor and two 650-Mbyte disks. All computational nodes can directly access the I/O nodes and the disks through the interconnection network. The data transfer rate between a single computational node and an I/O node is 1.5 Mbytes per second.

Like most other hypercubes, the iPSC/860 hypercube is not a stand-alone machine. It needs a host machine to serve as its interface to the programmers and outside network connection for program development and resource management. The host machine for an iPSC/860 hypercube is an Intel 301 microcomputer consisting of an Intel 80386/387 processor pair running at 16 MHz, with 8 Mbytes of RAM and a 300-Mbyte disk. The host is linked to a hypercube throngh communication channels.

Application programmers can use FORTRAN and C to develop codes on iPSC/860 hypercubes. The programming paradigm is to run codes on different computational nodes with each node working on part of the problem to be solved, and to collaborate the computation on different nodes by using explicit message exchanges.

3.2   Inter-processor Communications on iPSC/860

Since all nodes on a hypercube are highly connected, many different communication
patterns can be embedded into a hypercube with communication proximity preserved,
which means message exchange happens only between the directly connected nodes.
To achieve this, we first need to analyze the connection topology of a hypercube and
then develop tools for mapping computational tasks to processors.

3.2.1   Connection Topology

On a d-dimensional hypercube, each computational node has a unique node ID number which is a d-digit binary number. Two nodes on a hypercube will be directly
connected if and only if their binary d-digit ID numbers differ in only one bit. Fig.
3.3 shows the numbering and connection of the eight nodes on a three-dimensional

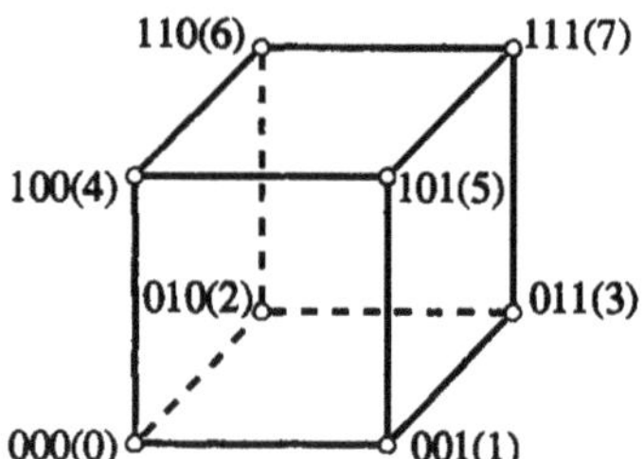

Figure 3.3: The numbering and connection of nodes on a three-
dimensional hypercube. The numbers in parentheses are
the decimal representations of node ID numbers.

hypercube. Both the binary and decimal node ID numbers are shown in the figure.
It is clear in Fig. 3.3 that each node is directly connected to three other nodes whose
binary ID numbers differ in only one bit with its own ID number.

In many application circumstances, for example in the domain decomposition
method, one often has a number of tasks, say T_0, T_1, $\cdots$, T_7, to be solved in parallel.
The communication pattern between these jobs is usually

$$T_0 \longleftrightarrow T_1 \longleftrightarrow T_2 \longleftrightarrow \cdots \longleftrightarrow T_7. \tag{3.1}$$

In other words, each task in this one-dimensional chain communicates with its nearest
neighbors. Given a hypercube of eight nodes and eight computational tasks, it seems

to be natural to assign task T_i to node P_i for parallel processing. While this task-to-node mapping is simple and straightforward, it does not preserve the proximity of communications between different tasks. For example, the pattern shown in (3.1) indicates that task T_3 needs to communicate with tasks T_2 and T_4. However, Fig. 3.3 shows that node P_3 is not directly connected to node P_4. Therefore, the communication has to be routed through intermediate nodes (for example nodes P_1 and P_5). To preserve the proximity of communications between different tasks, the binary reflected gray code can be used to map tasks T_i's to nodes P_i's. For a d-dimensional hypercube, the binary reflected gray code is a sequence of d-bit binary numbers such that any two successive numbers in the sequence differ in only one bit and all d-bit binary numbers are represented in the sequence. The trivial one-bit binary gray code is the sequence

$$G_1 = \{0, 1\}. \tag{3.2}$$

The d-bit gray code can be built recursively from the $(d-1)$-bit gray code by first prepending a 0 to each element in the $(d-1)$-bit gray code sequence, and then prepending a 1 to the same $(d-1)$-bit gray code sequence in a reversed order. The union of all those elements constitutes the d-bit gray code. Based on (3.2), the 2-bit gray code sequence is

$$G_2 = \{00, 01\} \oplus \{11, 10\} = \{00, 01, 11, 10\}, \tag{3.3}$$

and based on (3.3), the 3-bit gray code is

$$\begin{aligned} G_3 &= \{000, 001, 011, 010\} \oplus \{110, 111, 101, 100\} \\ &= \{000, 001, 011, 010, 110, 111, 101, 100\}. \end{aligned} \tag{3.4}$$

One can think of all the elements in G_3 as the node ID numbers of a three-dimensional hypercube. Since any two successive numbers in G_3 differ in only one bit, the sequence in G_3 gives a list of directly connected nodes. If we map the tasks T_i to the computational nodes based on the gray code, then the communication between two successive tasks will happen between directly connected computational nodes. For a three-dimensional hypercube, the mapping is

$$\begin{array}{c} \{T_0, T_1, T_2, T_3, T_4, T_5, T_6, T_7\} \\ \updownarrow \\ \{P_0, P_1, P_3, P_2, P_6, P_7, P_5, P_4\} \end{array} \tag{3.5}$$

where T_i's are computational tasks and P_i's are computational nodes. The task T_i needs to communicate with T_{i-1} and T_{i+1}, except for T_0 and T_7 which communicate with T_1 and T_6, respectively. The mapping in (3.5) shows the embedding of a one-dimensional chain into a three-dimensional hypercube. The idea can be generalized to embed a m-dimensional topology into a d-dimensional hypercube with $m < d$. Fig. 3.4(a) shows a set of 16 computational tasks that need to be mapped to a four-

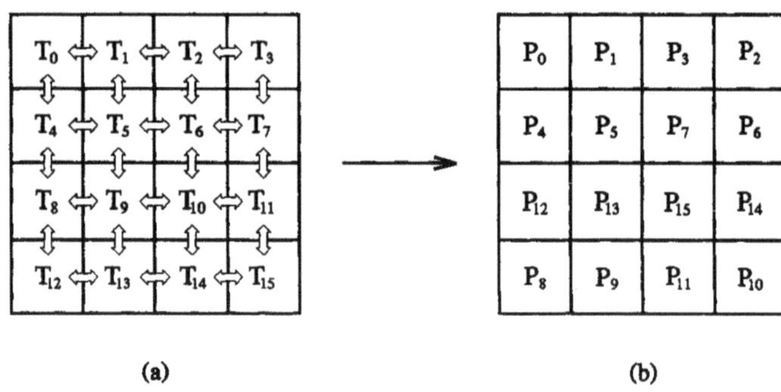

(a)   (b)

Figure 3.4: Mapping of computational tasks with a two-dimensional connection to a four-dimensional hypercube of 16 nodes. T_i's are computational tasks and P_i's are computational nodes on the hypercube. Arrows in (a) indicate communications.

dimensional hypercube for parallel execution. The tasks may correspond to a two-dimensional domain decomposition algorithm. A task T_i will in general communicate with the neighboring tasks in the east, west, north and south, except for the tasks on the boundary which communicate with two or three tasks. Fig. 3.4(b) shows the mapping of the tasks to computational nodes based on the gray code. It is easy to verify in Fig. 3.4(b) that any two neighboring tasks are mapped to two directly connected computational nodes, so that communications are all between the directly connected nodes. The gray code for this two-dimensional connection topology can be generated from the binary reflected gray code discussed before. The 4-bit gray code is

$$G_4 \;=\; \{0000, 0001, 0011, 0010, 0110, 0111, 0101, 0100,$$
$$1100, 1101, 1111, 1110, 1010, 1011, 1001, 1000\}.$$

If we divide G_4 into 4 successive segments as

$$\begin{aligned}
&\{0000, 0001, 0011, 0010\} \\
&\{0110, 0111, 0101, 0100\} \\
&\{1100, 1101, 1111, 1110\} \\
&\{1010, 1011, 1001, 1000\}
\end{aligned} \qquad (3.6)$$

and then reverse the order of numbers in the second and fourth segments to get

$$\begin{aligned}
&\{0000, 0001, 0011, 0010\} \\
&\{0100, 0101, 0111, 0110\} \\
&\{1100, 1101, 1111, 1110\} \\
&\{1000, 1001, 1011, 1010\},
\end{aligned} \qquad (3.7)$$

the decimal representation of (3.7) is

$$\begin{aligned}
&\{\ 0, \quad 1, \quad 3, \quad 2\} \\
&\{\ 4, \quad 5, \quad 7, \quad 6\} \\
&\{12, \quad 13, \quad 15, \quad 14\} \\
&\{\ 8, \quad 9, \quad 11, \quad 10\}
\end{aligned}$$

which is exactly the processor assignment shown in Fig. 3.4(b). It is obvious in (3.7) that every number in the table differs from its neighboring numbers (in both horizontal and vertical directions) by only one bit. For more general cases, if there are totally 2^d computational tasks T_i that form a two-dimensional $l \times m$ communication grid, we can map the 2^d computational tasks to a d-dimensional hypercube based on the gray code, so that all communications happen between the directly connected nodes. Assume $l = 2^{d_1}$, $m = 2^{d_2}$, and $d_1 + d_2 = d$, the d-bit gray code G_d should be cut into m segments with l numbers in each segment. If the segments are numbered 1, 2, $\cdots$ $m = 2^{d_2}$, then all the numbers in the even-numbered segments should be re-arranged in a reversed order, like what happened to row 2 and row 4 in (3.6). Each number in the final table, like that shown in (3.7), gives the ID number of a computational node that should be assigned to the computational task in that location.

If the computational tasks form a two-dimensional grid with dimensions $l' \times m'$ that are not powers of 2, a hypercube with $l \times m$ nodes must be used to preserve communication proximity, where l and m are the smallest numbers such that they are powers of 2 and $l \geq l'$, $m \geq m'$ respectively.

For example, if there are 9 computational tasks that form a 3×3 grid as shown in Fig. 3.5(a), then a hypercube of $4 \times 4 = 2^4 = 16$ nodes must be used to preserve the communication proximity. Only 9 out of the 16 available nodes are used in the

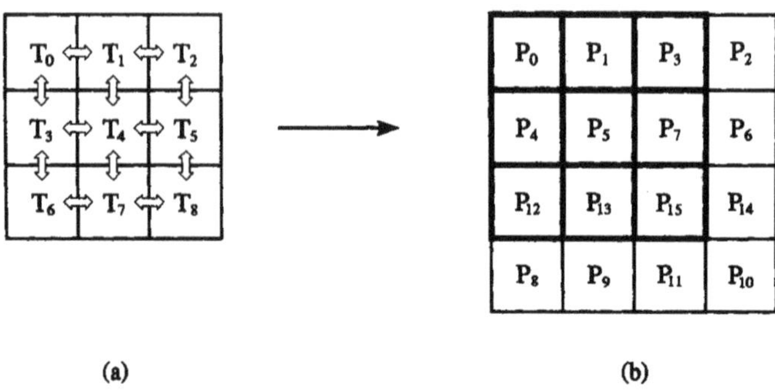

(a)   (b)

Figure 3.5: (a) Nine computational tasks form a 3 × 3 grid. Arrows
indicate communication. (b) Nine computational nodes of a
hypercube with 16 nodes are assigned to the computational
tasks. The communication proximity is preserved.

computation. Note, however, that the 9 computational nodes used are not successive
nodes numbered from 0 to 8.

For the computational tasks that form a three-dimensional grid, like that in three-
dimensional domain decomposition algorithms, the binary reflected gray code can
also be used to map the nodes on a hypercube to the computational tasks so that
communication proximity is preserved.

Assume that the computational tasks form a grid with dimensions $l \times m \times n$, where
$l = 2^{d_1}$, $m = 2^{d_2}$, and $n = 2^{d_3}$, a d-dimensional hypercube with $d = d_1 + d_2 + d_3$ is
needed for the computation. The three-dimensional computational grid can be cut
as the combination of n two-dimensional grids with dimensions $l \times m$. The d-bit gray
code can be generated as usual and then cut into $n = 2^{d_3}$ segments. If we number
these segments from 1 to n, then all numbers in the even numbered segments should
be re-arranged in a reversed order. Each of these n segments then corresponds to a
two-dimensional $l \times m$ grid, and the procedure described before for two-dimensional
computational grids can be applied here to assign computer nodes to computational
tasks in that two-dimensional grid.

Take a simple example shown in Fig. 3.6(a), the computational tasks form a
2 × 2 × 2 grid. Each computational task T_i needs to exchange messages with its
neighbors in all three directions. A hypercube of $8 = 2^3$ nodes is needed for the
computation. We can first cut the grid into two 2 × 2 grids as shown in Fig. 3.6(b).

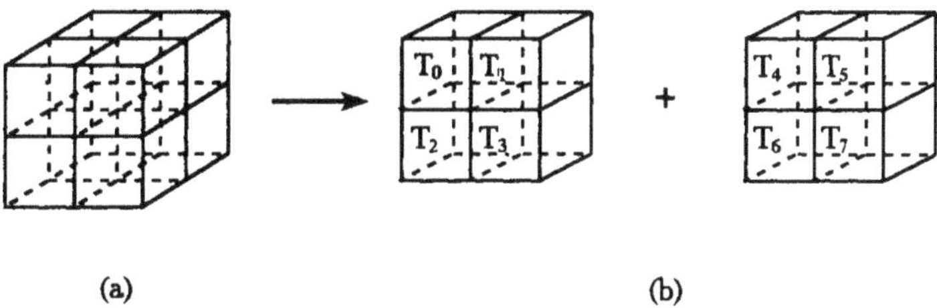

Figure 3.6: (a) Eight computational tasks form a $2 \times 2 \times 2$ grid. (b) The three-dimensional grid is cut into two two-dimensional grids. T_i represents computational tasks.

The 3-bit gray code

$$G_3 = \{000, 001, 011, 010, 110, 111, 101, 100\}$$

is then cut into two segments as

$$\{000, 001, 011, 010\} \quad \text{and} \quad \{110, 111, 101, 100\}.$$

The numbers in the second segments should be re-arranged in a reversed order as

$$\{000, 001, 011, 010\} \quad \text{and} \quad \{100, 101, 111, 110\}.$$

These two segments correspond to the two grids shown in Fig. 3.6(b). The procedure for the two-dimensional grids can then be used to generate the tables

$$\left\{ \begin{array}{l} 000, 001 \\ 011, 010 \end{array} \right\} \quad \text{and} \quad \left\{ \begin{array}{l} 100, 101 \\ 111, 110 \end{array} \right\}.$$

Re-arrange the numbers in the second rows of both tables, we get

$$\left\{ \begin{array}{l} 000, 001 \\ 010, 011 \end{array} \right\} \quad \text{and} \quad \left\{ \begin{array}{l} 100, 101 \\ 110, 111 \end{array} \right\}$$

which gives the node assignment for the two two-dimensional grids in Fig. 3.6(b). It is easy to verify from Fig. 3.7 that communication proximity is preserved in all three dimensions. It is a coincidence that the gray code task-to-node assignment is the same as the simple $T_i \rightarrow P_i$ assignment for this simple example.

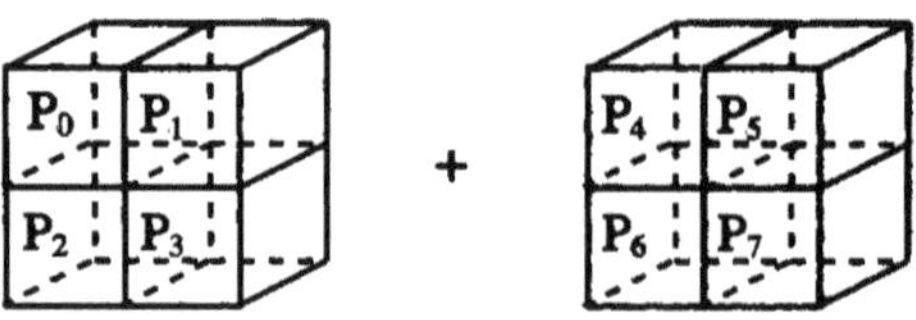

Figure 3.7: Node assignment based on gray code for the computational grid of Fig. 3.6(b).

In summary, the hypercube architecture has a highly connected topology that can accommodate many different computational problems and preserve the communication proximity. A programmer needs to study carefully the problem that is to be solved on hypercubes and sort out the communication patterns between different tasks, so that the computational nodes can be assigned to the tasks in such a way that communication proximity is preserved.

3.2.2   Inter-processor Communications

On the iPSC/860 system, the Direct Connect Module (DCM) on each processor provides a message passing mechanism for message exchanges. The time required to send a message of m bytes from one node to the nearest neighboring node is

$$T = \sigma + \beta m$$

where σ is the start-up time (latency) for the communication channel, β is the time required to send one byte of data after start-up. Each message is identified by an integer which is called message type on iPSC/860 systems. There are two message types depending on the values of the integer [96]. If the integer is between 1,073,741,824 to 1,999,999,999, it is called the *forced message type*. Otherwise, between 0 to 999,999,999, it is the *normal message type*. Most of the time, the normal message type is used to exchange messages between processors. A program running on one processor can send a message to and receive a message from another processor by calling subroutines csend and crecv, respectively. The calling sequence for these two subroutines are

```
call csend(mtype,buff,length,node_id,proc_id)
call crecv(mtype,buff,length)
```

where mtype is the message type, buff is a variable or an array containing the message to be sent or received, length is the number of bytes of the message, node_id is the ID number of the node to which the message is to be sent, and proc_id is the ID number of the process to which the message is to be sent. Note that the subroutine crecv picks up a right message solely based on the parameter mtype, not on from which node and process the message was sent. Therefore, it is very important to assign each message a unique message type number mtype.

When the subroutine csend or crecv is called at a computational node, the program execution will halt until the message has been sent to another node or received by the calling node. Therefore, they are called blocking communication subroutines.

There are also nonblocking versions of csend and crecv, called isend and irecv, for more effective communications. The reason for using non-blocking versions of communication subroutines is to overlap communications with computations. In many circumstances, a node that is sending a message to another processor has a lot of computations to do after sending the message. Since the communication on each node is handled by the DCM, the CPU can continue the computation once it activates the DCM for the communication, provided that the computation is independent of the message exchange. The situation for receiving a message is similar. If the message to be received has no effect on the current computation, the CPU can continue computation once it activates the DCM to receive the message. At a later point if one wants to make sure that the message has been sent to another node or has been received at the current processor, the subroutine "msgwait" can be called to synchronize the execution. If the message passing action has not completed yet, the "msgwait" call will put the program execution on hold until the message passing has finished.

Fig. 3.8 shows two processors P_1 and P_2 exchanging messages through their DCMs.

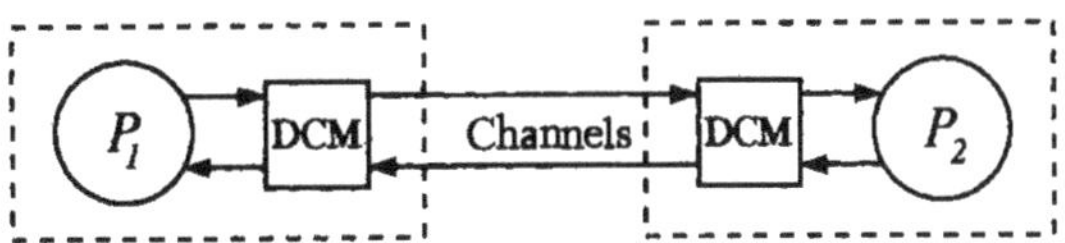

Figure **3.8**: A pair of processors exchanging messages through bi-direction channels.

A processor can accomplish message exchange by the following procedure:

Algorithm 3.1

1. get ready to receive the message data from the source processor by calling **irecv**

2. send the message data to the destination processor by calling **csend**

3. wait for message

If the normal message type is used, the execution of **csend** on one processor will go through the following steps [97]:

1. *The source processor establishes a path from the source processor to the destination processor by sending a probe;*

2. *The destination processor sends an acknowledgement to the source processor after receiving the probe signaling that it is ready for the message;*

3. *After the source processor receives the acknowledgement, message transmission begins through the established path;*

4. *When all message data arrive at the destination processor, the connection path is released.*

The four-step message passing process is performed by the DCM which is attached to each processor of a hypercube. The disadvantage of the four-step message passing process is that the acknowledgement step usually prevents messages from being sent simultaneously by the two communicating processors through the bi-directional channel. Therefore, the effective bandwidth is only half of the designed capacity. On iPSC/860 hypercubes, a processor can receive a message at the rate of 2.8Mbytes/second and send a message at the same rate simultaneously. Therefore, the designed ideal transmission rate between two processors is 5.6Mbytes/second.

However, if two processors P_1 and P_2 exchange their messages through the bi-directional channel using algorithm 3.1 with the normal message type, as shown in Fig.

3.8, it is more likely that one processor will execute the **csend** earlier than the other, since the program executions on the two processors can not be synchronized perfectly. If P_1 executes **csend** earlier than does P_2, and P_2 has sent acknowledgement back to P_1, then P_1 will start sending message data to P_2. When P_2's probe signal arrives, P_1 can not send acknowledgement back to P_2 until it finishes sending the message data, because there is only one forward channel form P_1 to P_2. Thus, processor P_2 can not start sending the message and has to wait until P_1 finishes sending the message and then returns the acknowledgement. Therefore processors P_1 and P_2 can not send messages to each other simultaneously through the bi-directional channels using the normal message type, except in the very unlikely event when the two **csends** are perfectly synchronized. The best observed exchange rate in the numerical experiments is close to only 2.8Mbytes/second.

If a message is labeled "forced type", the DCM will send the message directly to the destination processor without going through the four steps listed above. The one-step message passing process has no prelude of probing path and waiting for acknowledgement. If the receiving processor has not allocated storage for the incoming message when the message arrives, the message will be dumped and lost. This is different from the four-step message passing process which has a mechanism to guarantee that the message is received by the destination processor: The message will be put into the system buffer if the receiving processor has not allocated memory for the message. Hence, with the forced message type, we have to carefully schedule the algorithm to ensure that the receiving processor gets ready for the incoming message before the other processor sends message data using the forced message type. More detailed discussions on forced message type will be given in Section 3.4.

3.3 Communication Analysis for Domain Decomposition Method

Domain decomposition algorithms [28, 68, 69, 105] are very popular for solving PDEs because it introduces natural parallelism on parallel computers. The basic idea is to decompose the original spatial domain where the problem is defined into subdomains and assign different subdomains to different processors for parallel processing. The correct physical relations between subdomains are retained during the solution process by inter-processor communications. The efficiency of an algorithm on a parallel computer depends very much on the implementation details of the algorithm [69], like how the domain is decomposed and how the inter-processor communication is scheduled.

For example, a two-dimensional rectangular domain can be decomposed into strips
as shown in Fig. 3.9(a), or into patches as shown in Fig. 3.9(b). The scheme in Fig.

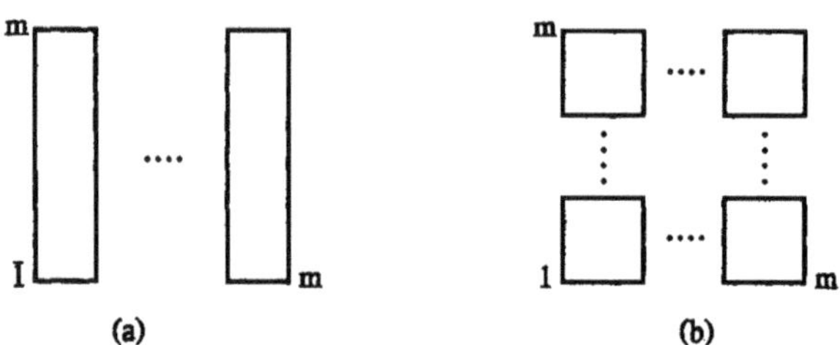

Figure 3.9: Two domain decomposition schemes for a two-dimensional
domain

3.9(a) corresponds to decomposing the domain in the horizontal dimension only, while
the scheme in Fig. 3.9(b) corresponds to decomposing the domain in two dimensions,
both the vertical and the horizontal dimensions. There are even more possibilities for
three-dimensional domains.

A spatial domain can be decomposed into subdomains in many different ways.
But most of the time, the domains are decomposed into subdomains with simple and
regular geometries for simplicity of algorithm implementations. The domains discussed here are either rectangular or cubic domains. For more complicated domains,
numerical grid generation schemes can be used to map the complex physical domains
to the regular computational domains [172].

3.3.1 Two-Dimensional Domains

We first discuss two ways to decompose a rectangular two-dimensional $m \times m$ domain, i.e. the strip decomposition (D_{21} as shown in Fig. 3.9(a)) and the patch
decomposition (D_{22} as shown in Fig. 3.9(b)).

Assume that a subdomain needs to exchange a message with all adjacent subdomains, and that the number of single precision data in a message is equal to the
number of grid points on the boundary of the subdomain. For the two-dimensional
$m \times m$ domain in Fig. 3.9, with the strip decomposition scheme each subdomain needs
to exchange a message of m single precision data with two neighboring subdomains.
The communication time on an Intel iPSC/860 hypercube will be

$$T_{D_{21}} = 2(\sigma + 4\beta m) \tag{3.8}$$

where σ is the communication start-up time and β is the time required to send one byte of data. The communication (send and receive) subroutines on iPSC/860 require that the data to be exchanged locate in contiguous memory units. If the grid data is stored in an array $A(m, l)$, where $l = m/P$ and P is the number of processors, the data corresponding to the two boundaries of a subdomain are stored in $A(1 : m, 1)$ and $A(1 : m, l)$ respectively, which are the first and the last columns of the array respectively. With FORTRAN, data elements in an array are stored by columns, therefore the two data sets to be exchanged are located in the contiguous memory units and can be exchanged directly. If other programming languages are used which map array elements to memory units by rows, the grid data of the subdomain should be transposed and stored in $A(l, m)$ so that the data to be exchanged are in the contiguous memory locations.

The disadvantage of this decomposition scheme is that the length of messages to be exchanged does not decrease as the number of processors increases. The message length is always m regardless how many processors are used in the computation. On the other hand, the amount of numerical computation is proportional to the number of grid points m^2/P in a subdomain, which decreases as the number of processors P increases. If the Jacobi iteration is used to solve the Laplace equation, the amount of numerical computation will be $4m^2/P$ for each iteration. Therefore the communication/computation ratio is

$$r_{21} = \frac{2P(\sigma + 4\beta m)}{4m^2\tau} = \frac{P(\sigma + 4\beta m)}{2m^2\tau}$$

which is proportional to the number of processors P used for the computation. τ is the time required for a floating point operation.

For decomposition scheme D_{22}, assume $\sqrt{P}$ is an integer and m is divisible by $\sqrt{P}$. The processors can be arranged as a $\sqrt{P} \times \sqrt{P}$ grid and a subdomain will have dimensions $\frac{m}{\sqrt{P}} \times \frac{m}{\sqrt{P}}$. A subdomain needs to communicate with four adjacent subdomains, except the subdomains on the boundary of the original domain which communicate with two or three subdomains. The message length in each message exchange is $\frac{m}{\sqrt{P}}$. The communication time is

$$T_{D_{22}} = 4(\sigma + 4\beta \frac{m}{\sqrt{P}}).$$

In contrast to $T_{D_{21}}$, we can see that $T_{D_{22}}$ decreases as the number of processors increases. The amount of numerical computation is the same as in scheme D_{21}, since a strip and a patch have the same number of grid points. Therefore the communication/computation ratio of scheme D_{22} is

$$r_{22} = \frac{4P(\sigma + 4\beta \frac{m}{\sqrt{P}})}{4m^2\tau} = \frac{\sigma P + 4\beta m\sqrt{P}}{m^2\tau}.$$

It can be seen that one term which corresponds to the start-up time in r_{22} increases proportionally to P. For long messages, the communication time is dominated by the second term $4\beta m\sqrt{P}$ which is a slowly increasing function.

Fig. 3.10 shows the curves representing r_{21} and r_{22} vs. the number of processors

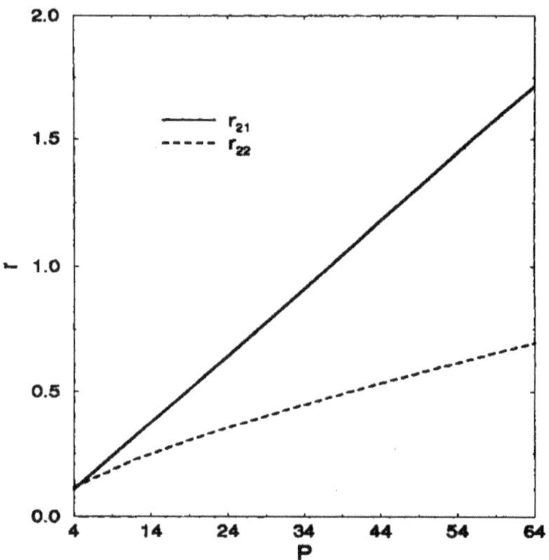

Figure 3.10: communication/computation ratios r_{21} and r_{22} vs. the number of processors P for decomposition schemes D_{21} and D_{22}.

used in the computation. The parameter values are $m = 512$, $\sigma = 100$ (μs), $\beta = 0.5$ (μs), and $\tau = 0.08$ (μs). The machine parameters σ, β, and τ are based on numerical experiments like those done in [78], not on the peak performance values. From Fig. 3.10 it is clear that scheme D_{22} has a lower communication/computation ratio r_{22} which increases much more slowly than does r_{21}.

The disadvantage with decomposition scheme D_{22} is that an interior subdomain needs to communicate with four adjacent subdomains, as opposed to two subdomains with D_{21}. If the grid data of a subdomain is stored in an array $A(l, l)$ where $l = \frac{m}{\sqrt{P}}$, then the first and the last columns and the first and the last rows of A need to be exchanged with the four adjacent subdomains respectively. As discussed before, the first and last columns of A can be exchanged directly by calling communication subroutines on the iPSC/860 since they are stored in the contiguous memory locations. The data items of the first and last rows of A, however, are not stored in the contiguous memory units using FORTRAN. Therefore, they have to be packed into one-dimensional arrays before communication, typically using the following statements:

$$
\begin{array}{l}
\texttt{do i=1,}l \\
\quad\texttt{buff1(i)=A(1,i)} \\
\quad\texttt{buff2(i)=A(}l\texttt{,i)} \\
\texttt{end do}
\end{array}
\tag{3.9}
$$

To avoid doing this, the programmer must store the boundary data separately from the interior data of the subdomain.

The curves in **Fig.** 3.11 show the timing results of message exchange for a 512 ×

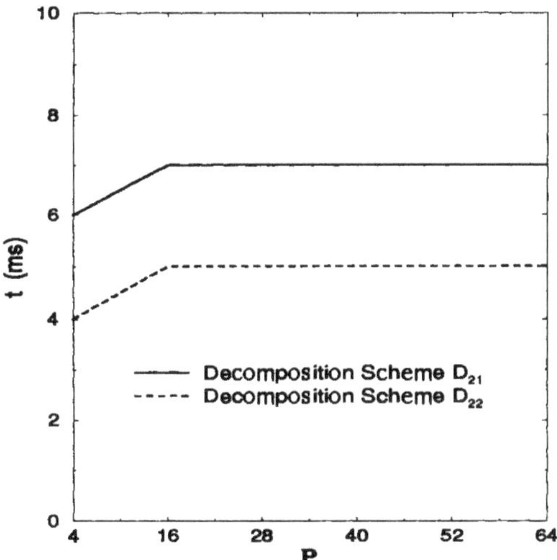

Figure 3.11: Timing curves of decomposition schemes D_{21} and D_{22} for exchanging messages of a 512 × 512 domain.

512 two-dimensional domain with schemes D_{21} and D_{22} respectively. With scheme D_{21}, processors are configured as a chain and each processor needs to communicate with two processors a message of 512 single precision data, except the first and last processors in the chain which communicate with only one processor. When scheme D_{22} is used, the P processors are arranged as a $\sqrt{P} \times \sqrt{P}$ grid, i.e. 2×2 (4 processors), 4×4 (16 processors), and 8×8 (64 processors) grids. In the more general case when $\sqrt{P}$ is not an integer, the P processors can be configured as a $n_t \times n_2$ grid with $P = n_1 \times n_2$. The message length is 256, 128 and 64 single precision data for 4, 16 and 64 processors, respectively.

Fig. 3.11 shows that scheme D_{22} takes less time to exchange messages between subdomains. However, the time used by scheme D_{22} does not decrease as the number of processors increases. In the case of 4 processors arranged as a 2×2 grid, each processor needs to communicate with actually only two processors with a message length of 256 words. When 16 processors are used (arranged as a 4×4 grid), the communication time is dominated by the processors that have to exchange messages with 4 adjacent processors. Although the message length is cut to 128, the reduction of the communication time from the shorter message length is not enough to offset the increase of the start-up times of 4 communications, therefore the total communication time actually went up in the case of 16 processors.

The curves in Fig. 3.12 have the same meaning as that in Fig. 3.11, except that

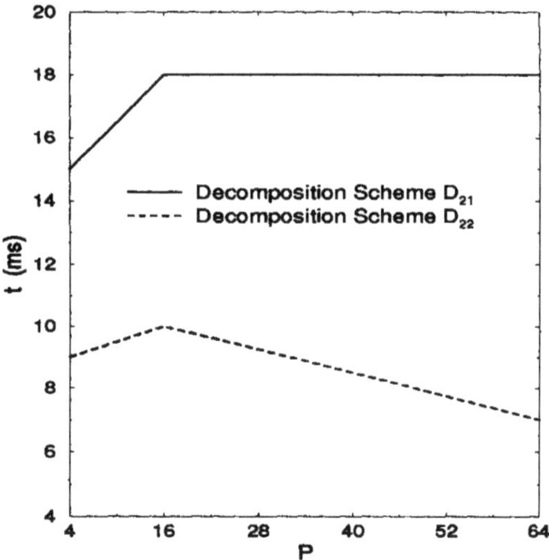

Figure 3.12: Timing curves of decomposition schemes D_{21} and D_{22} for exchanging messages of a 2048×2048 domain.

a much larger domain is used (2048×2048). The decomposition scheme D_{22} uses much less communication time than does scheme D_{21}. For the same reason discussed for Fig. 3.11, it takes more time for 16 processors to exchange messages than for 4 processors. However, since a much larger domain is involved here, the reduction of the communication time from shorter message length by using more processors eventually offsets the additional communication overhead incurred as the number of processors increases. Therefore, the total communication time decreases as the number of processors increases.

3.3.2 Three-Dimensional Domains

For a $m \times m \times m$ three-dimensional domain, three decomposition schemes are discussed here, i.e. the slice decomposition (D_{31} shown in Fig. 3.13), the column decomposition

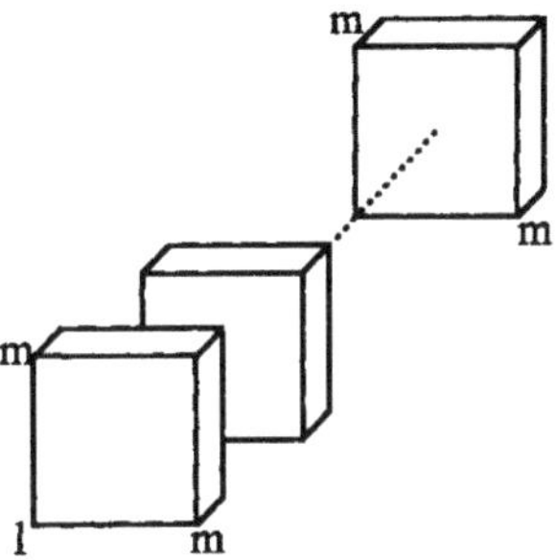

Figure 3.13: Slice decomposition (D_{31}) of a three-dimensional $m \times m \times m$ domain

(D_{32} shown in Fig. 3.14), and the block decomposition (D_{33} shown in Fig. 3.15).

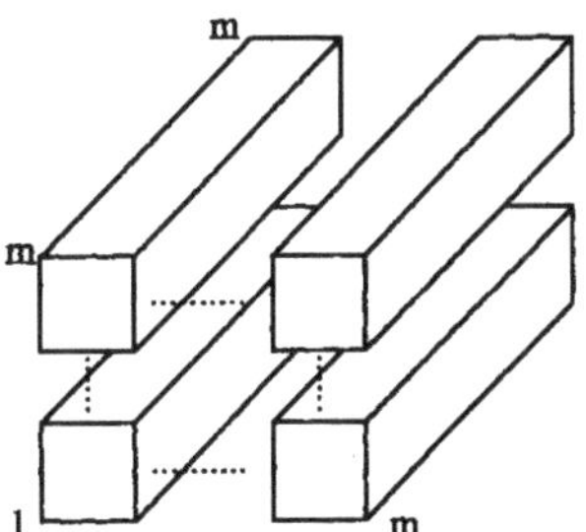

Figure 3.14: Column decomposition (D_{32}) of a three-dimensional $m \times m \times m$ domain

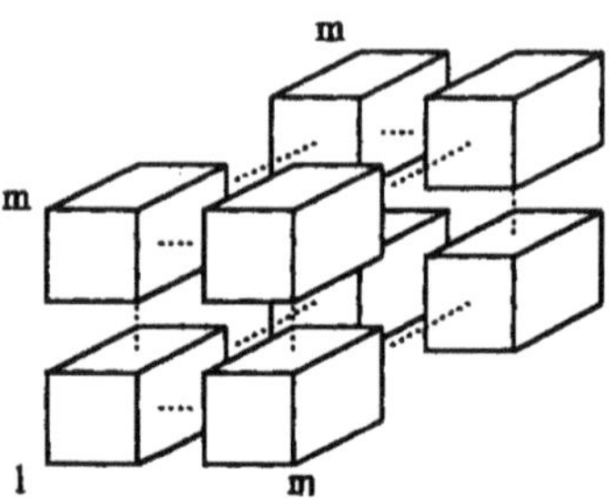

Figure 3.15: Block decomposition (D_{33}) of a three-dimensional
$m \times m \times m$ domain

With scheme D_{31}, the situation is very much the same as with scheme D_{21} for the
case of a two-dimensional domain, except that the message length is now m^2 and the
amount of computation is proportional to $6\frac{m^3}{P}$. The total communication time is

$$T_{D_{31}} = 2(\sigma + 4\beta m^2)$$

and the communication/computation ratio is

$$r_{31} = \frac{2(\sigma + 4\beta m^2)}{6\frac{m^3}{P}\tau} = \frac{P(\sigma + 4\beta m^2)}{3m^3\tau}.$$

If the grid data of the subdomain are stored in an array $A(m, m, l)$ with $l = m/P$, the
boundary data to be exchanged are all in contiguous memory locations and can be
transferred directly. The ratio r_{31} increases proportionally to the number of processors
as does r_{21}.

If the domain is decomposed in two dimensions and the third dimension remains
intact, as shown in Fig. 3.14, each processor needs to exchange with four adjacent
columns in general, except the processors on the boundary. If the P processors are
arranged as a $\sqrt{P} \times \sqrt{P}$ grid, then the message length of each communication is $\frac{m^2}{\sqrt{P}}$.
The total communication time is then

$$T_{D_{32}} = 4(\sigma + 4\beta \frac{m^2}{\sqrt{P}})$$

and the communication/computation ratio is

$$r_{32} = \frac{4(\sigma + 4\beta \frac{m^2}{\sqrt{P}})}{6\frac{m^3}{P}\tau} = \frac{2\sigma P + 8\beta m^2 \sqrt{P}}{3m^3\tau}.$$

This is similar to the case when scheme D_{22} is applied to a two-dimensional domain. The term in r_{32} corresponding to the start-up time grows proportionally to P, while the other term grows as a function of $\sqrt{P}$.

Fig. 3.15 shows the block decomposition scheme D_{33} in which all three dimensions of the original domain is decomposed. Assume that the processors are arranged as a $P^{\frac{1}{3}} \times P^{\frac{1}{3}} \times P^{\frac{1}{3}}$ grid, the subdomains then have dimensions $\frac{m}{\sqrt[3]{P}} \times \frac{m}{\sqrt[3]{P}} \times \frac{m}{\sqrt[3]{P}}$. A processor holding an interior subdomain needs to communicate a message of length $\frac{m^2}{\sqrt[3]{P^2}}$ with 6 processors holding adjacent blocks. The total communication time is then

$$T_{D_{33}} = 6(\sigma + 4\beta \frac{m^2}{\sqrt[3]{P^2}})$$

and the communication/computation ratio is

$$r_{33} = \frac{6(\sigma + 4\beta \frac{m^2}{\sqrt[3]{P^2}})}{6\frac{m^3}{P}\tau} = \frac{\sigma P + 4\beta m^2 \sqrt[3]{P}}{m^3 \tau}.$$

We can see that the second term in r_{33} grows as a cubic root function of P.

Fig. 3.16 shows the curves of r_{31}, r_{32} and r_{33} vs. the number of processors P.

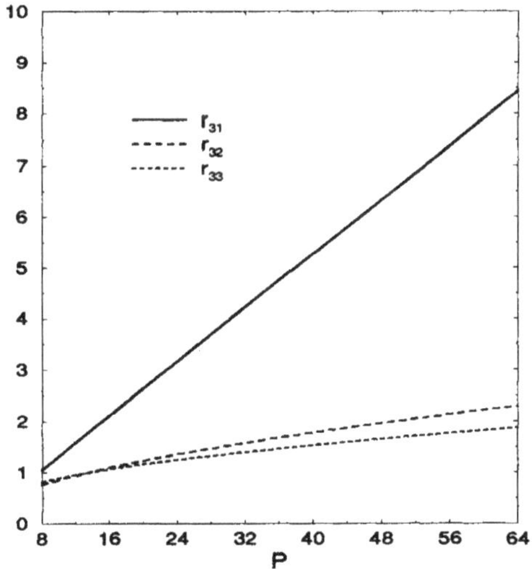

Figure 3.16: Communication/computation ratios vs. the number of processors P for decomposition schemes D_{31}, D_{32} and D_{33}.

It is clear that scheme D_{33} has the lowest communication/computation ratio as P increases.

Similar to the situation when scheme D_{22} is applied to a two-dimensional domain, with schemes D_{32} and D_{33}, some boundary data to be exchanged must be packed into a one-dimensional array before communication, as expressed by (3.9). The influence of this data packing is hard to quantify analytically. Figs. 3.17 and 3.18 show the timing

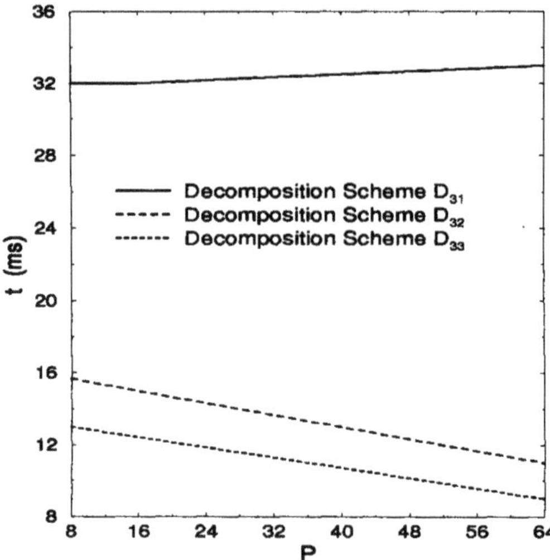

Figure 3.17: Timing curves of decomposition schemes D_{31}, D_{32} and D_{33} for exchanging messages of a $64 \times 64 \times 64$ domain.

curves of the three decomposition schemes for a $64 \times 64 \times 64$ and a $128 \times 128 \times 128$ domains. For scheme D_{32}, processors are arranged as 4×4 and 8×8 grids. For scheme D_{33}, processors are arranged as $2 \times 2 \times 2$ and $4 \times 4 \times 4$ grids respectively. It can be seen from Figs. 3.17 and 3.18 that scheme D_{33} takes the least amount of time to exchange boundary data of the subdomains, even though a processor has to communicate with six processors and pack some boundary data.

From the analysis given here, it is clear that for large spatial domains, decomposing in all dimensions generates subdomains with better communication/computation ratios, as indicated by r_{22} and r_{33} in Figs. 3.10 and 3.16, respectively. An intuitive explanation for this is that the communication/computation ratio is actually proportional to the surface/volume (or circumference/area) ratio of a subdomain, and decomposing an $m \times m \times m$ spatial domain in all three dimensions will generate subdomains with better surface/volume ratio. However, if the original spatial domain is not a cube with the same sizes in all dimensions, it may not be the best to de-

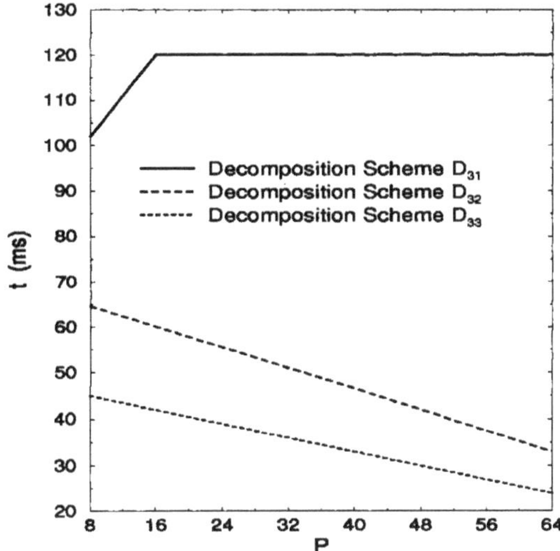

Figure 3.18: Timing curves of decomposition schemes D_{31}, D_{32} and D_{33} for exchanging messages of a $128 \times 128 \times 128$ domain.

compose the domain in all dimensions. The decomposition of a spatial domain into subdomains should be carried out in such a way as to generate subdomains that have as nearly equal sizes in all dimensions as possible.

3.4 Bandwidth Improvement using Forced Message Type

As discussed in Section 3.2, message passing using forced message type is more straightforward and can be more efficient than using the normal message type. However, there is the risk of losing the message being transferred if the receiving processor is not ready for the incoming message. A node is considered to be ready for receiving a message if a message receiving subroutine (either **crecv** or **irecv**) has been called and a proper message buffer (memory) has been allocated. Therefore, to ensure that the message sent by a processor using the forced message type is received, the receiving processor must call the **crecv** or **irecv** subroutine before the incoming message arrives. To overlap communications with computations, the receiving processor should call the non-blocking **irecv** whenever possible to allocate storage for the incoming message.

3.4.1   Message Passing Using Forced Message Type

We now discuss the situation when two nodes exchange messages. Since it is unlikely that both processors can call **irecv** at the same time, each processor must signal the other processor that it has called the **irecv** and storage is allocated for the incoming message. This is accomplished by exchanging a zero-byte message between two processors. After two processors receive the zero-byte messages, which means both processors are prepared for the incoming messages, they then start sending message data directly to the destinations using the forced message type. In this way, both processors can send message data simultaneously without waiting for the acknowledgement from the other processor. The throughput was observed to approach 5.6 Mbytes/second for long messages in numerical experiments [199].

The algorithm for exchanging a message between a pair of processors using the forced message type can be described as [97]:

Algorithm 3.2

1. post an **irecv** to get ready to receive the message data

2. send a zero-byte signal to the destination processor by calling **csend**

3. receive a zero-byte signal by calling **crecv**

4. call **csend** to send message data using the forced message type

5. wait for message

The purpose of steps 2 and 3 is to make sure that the receiving processors have allocated storage for the incoming messages. The **csend** in step 4 using the forced message type puts the message directly to the destination processor without going through the process of sending a probe and waiting for the acknowledgement. This makes it possible for two processors to exchange messages simultaneously. However, in algorithm 3.2, a processor sends then receives a zero-byte message (steps 2 and 3) before the exchange of the real messages starts, which causes additional communication overhead. This additional overhead can usually be offset by the improved communication speed of the real message data in step 4, especially for long messages.

Based on algorithm 3.2 which is for the message exchange between a pair of processors, we will develop, in the next two sections, algorithms for message exchange in a one-dimensional chain and a two-dimensional mesh, respectively.

3.4.2 Application: One-Dimensional Domain Decomposition

For a given space domain, there are many different ways [68, 69, 105] to decompose it. The simplest scheme is to decompose a domain in one dimension only, as shown in Fig. 3.19, where a two-dimensional domain is decomposed into strips. A three-dimensional

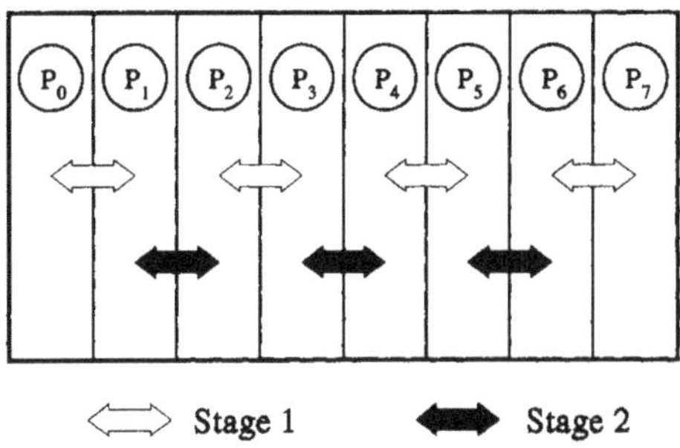

Figure 3.19: A two-dimensional domain decomposed into strips and the two stages of message exchanges.

domain can be decomposed into slices in a similar way. Typically, solution values after each iteration on the boundaries of a subdomain need to be exchanged with the adjacent subdomains. For example, if the Jacobi iterative algorithm is used to solve the Laplace equation, values on the boundaries of a strip need to be exchanged with the neighboring strips after each Jacobi iteration. From Fig. 3.19, we can see that each processor will exchange a message with its left and right neighbors except the first one which has no predecessor and the last one which has no successor. If processor P_i is assigned a strip, it needs to exchange messages with P_{i-1} and P_{i+1}, $i = 1, 2, \cdots, n - 2$, except P_0 and P_{n-1} which exchange a message with P_1 and P_{n-2}, respectively. Here n is the number of processors. A straightforward communication scheme for processor P_i without using the forced message type can be described as:

Algorithm 3.3

if $(i - 1 \geq 0)$ **then**
    receive a message from P_{i-1} by calling
    irecv send a message to P_{i-1} by calling
    csend wait for messages
end if
if $(i + 1 \leq n - 1)$ **then**
    receive a message from P_{i+1} by calling
    irecv send a message to P_{i+1} by calling
    csend wait for messages
end if

As discussed in the previous section, processors must be grouped into pairs to take advantage of the bi-directional channels using the forced message type. For the one-dimensional chain in Fig. 3.19, message exchange between processors can be done in two stages. In the first stage, processors connected by the blank arrows communicate with each other pairwisely; in the second stage, processors connected by the dark arrows communicate with each other pairwisely.

Assuming that the number of processors is even, then the communication scheme using the forced message type for processor P_i can be written as:

Algorithm 3.4

Stage 1 (blank arrows)
if (processor number i is even) **then**
    exchange with processor P_{i+1} using algorithm 3.2
else
    exchange with processor P_{i-1} using algorithm 3.2
end if

Stage 2 (dark arrows)
if (processor number i is odd) **then**
    exchange with processor P_{i+1} using algorithm 3.2
else
    exchange with processor P_{i-1} using algorithm 3.2
end if

Table 3.1 shows the timing results (in milliseconds) of interprocessor communica-

Table 3.1: Timing results of message exchange in a chain of processors with and without the forced message type.

L \ n	2		4		8		16		32	
	T_f	T_{nf}	T_f	T_{nf}	T_f	T_{nf}	T_f	T_{nf}	T_f	T_{nf}
256	3	3	4	4	4	4	3	5	3	6
512	3	4	4	5	5	6	4	7	5	7
1024	4	6	5	8	6	9	6	11	7	11
2048	5	8	8	14	9	15	9	18	9	19
10240	16	34	32	63	32	64	33	77	33	79

n: number of processors in the chain,
L: message length (single precision),
T_f : time using forced message type (ms),
T_{nf} : time without using forced message type (ms).

tions on a one-dimensional chain using algorithm 3.3 and 3.4 respectively, n is the number of processors in the chain, and L is the length of the messages exchanged.

It is obvious from Table 3.1 that the speed of communication using forced message type is faster than that without using the forced message type, especially for messages long enough so that the time spent on preparation for data exchange can be ignored. We can see that when $L = 10240$, $T_f < \frac{1}{2}T_{nf}$ for all values of n, which shows that the forced message type can take advantage of bi-directional channels, while using the normal message type can utilize at most half of the designed bandwidth. When $n = 2$, a processor has only one neighbor, therefore, the communication time is only half of that for n greater than 2. When forced message type is used, the communication times are about the same for $n = 4, 8, 16$ and 32, since all pairs of processors can exchange messages concurrently in the two stages. However, with algorithm 3.3, one processor deals with two neighbors at the same time. Therefore, the process of probing path and returning acknowledgement is more complicated, which causes the increase of communication time T_{nf} as the number of processors increases.

3.4.3   Application: Two-Dimensional Domain Decomposition

If a space domain is decomposed in two dimensions, message exchange will also be considered in two dimensions. Fig. 3.20 represents a domain decomposed into 16 subdomains. A straightforward message exchange scheme for processor P_j can be given as:

Algorithm 3.5

if (P_j has an east neighbor P_e) **then**
   receive a message from P_e
   send a message to P_e
   wait for messages
end if
 if (P_j has a west neighbor P_w) **then**
   receive a message from P_w
   send a message to P_w
   wait for messages
end if
 if (P_j has a north neighbor P_n) **then**
   receive a message from P_n
   send a message to P_n
   wait for messages
end if
 if (P_j has a south neighbor P_s) **then**
   receive a message from P_s
   send a message to P_s
   wait for messages
end if

The message exchange scheme with forced message type should proceed in four stages, i.e. two for each dimension. Note for each row in Fig. 3.20(a) and each column in Fig. 3.20(b), the message exchange process is exactly the same as that in the situation of a chain in one-dimensional case as given in Fig. 3.19. Therefore, we can apply algorithm 3.4 to the two-dimensional communication algorithm using the forced message type:

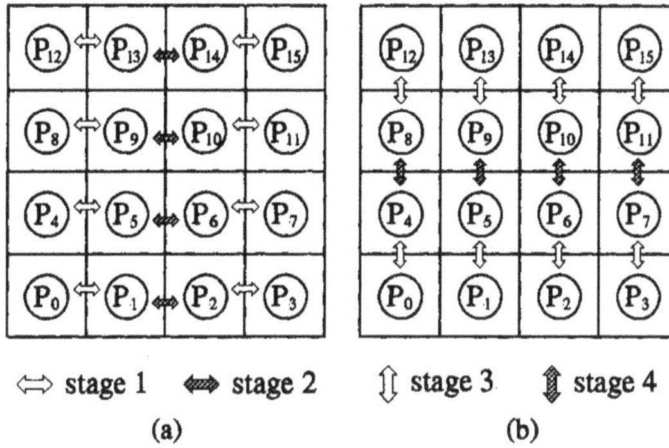

Figure 3.20: A two-dimensional domain decomposed into patchs with four stages of message exchanges.

Algorithm 3.6

1. all processors connected by horizontal arrows in Fig. 3.20(a) exchange messages pairwisely using algorithm 3.4 which has two stages indicated by the blank and dark arrows.

2. all processors connected by vertical arrows in Fig. 3.20(b) exchange messages pairwisely using algorithm 3.4 which has two stages indicated by the blank and dark arrows.

Table 3.2 shows the timing results for exchanging a message between neighboring processors with the normal and forced message types respectively. L is the length of the messages exchanged, and all processors are arranged as a two-dimensional grid with dimensions $n_1 \times n_2$.

Table 3.2 demonstrates similar results as those in Table 3.1. In the 2×2 case, each processor has only two neighbors. In the cases of 2×8 and 2×16, each processor has up to three neighbors. In these two cases, the communication process with forced

Table 3.2: Timing results of message exchange in a 2-D mesh of processors with and without the forced message type. (ms)

L \ $n_1 \times n_2$	2×2		2×8		2×16		4×4		4×8		8×8	
	T_f	T_{nf}	T_f	T_{nf}	T_f	T_{nf}	T_f	T_{nf}	T_f	T_{nf}	T_f	T_{nf}
64	3	3	3	3	3	4	3	4	4	5	6	6
256	3	3	4	5	4	6	5	7	5	7	7	9
512	4	5	5	8	6	8	6	10	6	10	8	12
1024	5	8	7	13	8	14	9	16	10	17	10	18
2048	8	15	12	23	12	24	15	28	15	34	17	35
4096	14	26	20	44	21	46	27	51	27	62	29	65

$n_1 \times n_2$: dimension of the processor mesh,
L: message length (single precision),
T_f : time using forced message type (ms),
T_{nf}: time without using forced message type (ms).

message type executes only three stages, one on the vertical dimension and two for the horizontal dimension.

For the cases of 4×4, 4×8, and 8×8, the communication process using the forced message type takes four stages. In each stage, all processors exchange messages pairwisely at the same time. Therefore, we see from the table that as the number of processors increases from $4 \times 4 = 16$ to $8 \times 8 = 64$, the times used to finish the message exchanges between all processors are about the same, particularly for long messages. But for the communication process without using the forced message type, the times increase more significantly as the number of processors increases. This is due to the same reason discussed at the end of section 3.4.2.

On a three-dimensional space domain decomposed into blocks in each dimension, a processor holding one block will communicate with six processors around itself. The process of communication using the forced message type is similar to that given in algorithm 3.6, except that the process should proceed in six stages instead of four.

3.5   KSR-1 Parallel Computers

The architecture of the KSR-1 parallel computer has been briefly discussed in Section 1.2.3. The main features of this architecture include distributed physical memory which makes the system scalable to a large number of processors, and a shared address space which provides users a shared-memory-like programming environment. Early system studies and performance evaluations are given in [14, 46, 152], and results of solving some application problems are reported in [53, 165, 202, 203].

Fig. 3.21 shows the architecture of KSR-1 parallel computers [108]. Each processor

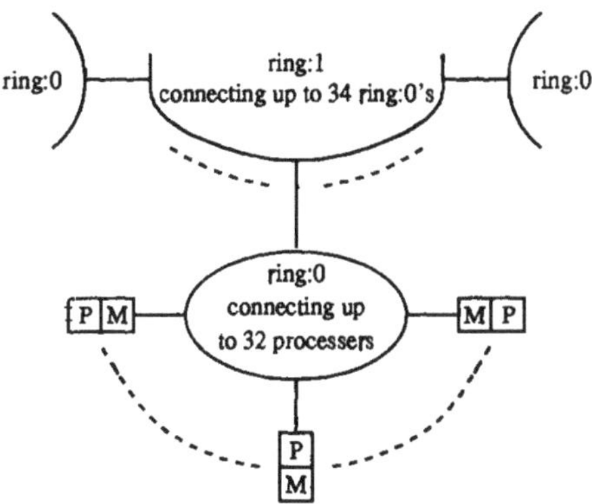

Figure 3.21: Configuration of the KSR-1 parallel computers. P: processor, M: 32 Mbytes of local memory

on a KSR-1 has 32 Mbytes of local memory. The CPU is a super-scalar processor with a peak performance of 40 MFlops in double precision. Different processors are organized as different rings. The local ring (ring:0) can connect up to 32 processors, and a higher level ring of rings (ring:1) can connect up to 34 local rings with a maximum of 1088 processors.

If a data element that is not located in the local memory of processor P_i is needed at that processor, the local search engine (SE:0) will search the processors in the local ring (ring:0) and try to get a copy of the data for processor P_i. If the search engine SE:0 can not locate the data element within the local ring, the request will be passed to the search engine at the next level (SE:1) to locate the data. This is done automatically by a hierarchy of search engines connected in a fat-tree-like structure [108, 118].

The memory hierarchy of the KSR-1 is shown in Fig. 3.22. Each processor has 512 Kbytes of fast *subcache* which is similar to the normal cache on other parallel computers. This subcache is divided into two equal parts as instruction subcache and

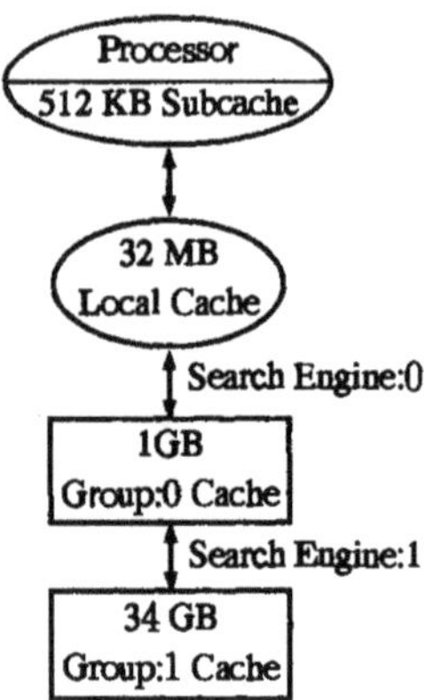

Figure 3.22: Memory hierarchy of KSR-1.

data subcache respectively. The 32 Mbytes of local memory on each processor is called *local cache*. A local ring (ring:0) with up to 32 processors can have 1 Gbytes of total local cache which is called *Group:0 cache*. Access to the Group:0 cache is provided by Search Engine:0. A higher level ring of rings (ring:1) connects up to 34 local rings with 34 Gbytes of total local cache (*Group:1 cache*). Access to the Group:1 cache is provided by Search Engine:1. The whole memory hierarchy is defined as ALLCACHE memory by the Kendall Square Research. Access by a processor to the ALLCACHE memory system is accomplished by going through different Search Engines as shown in Fig. 3.22. The latencies for different memory locations [107] are

1. 2 cycles for *subcache*,

2. 18 cycles for *local cache*,

3. 75 cycles for *Group:0 cache*,

4. 600 cycles for *Group:1 cache*.

3.6   Automatic, Semi-automatic and Manual Parallelization

In contrast to distributed memory parallel computers on which the explicit use of message passing subroutines is needed to coordinate the solution process, there is no explicit message passing involved in the program development on KSR-1 parallel computers. Therefore, the programming paradigm is different on KSR-1 from that on distributed memory parallel computers like iPSC/860 hypercubes. The most important concept in KSR-1 programming is the use of *"pthread"*. A pthread is a sequential flow of control within a process that cooperates with other pthreads to solve a problem. If we think of processors on KSR-1 as workbenches, then the pthreads are the workers who can work at different workbenches and coordinate with each other to make a product. If there are more workers then the number of workbenches, several workers may share one workbench. Also, a worker may switch to a different workbench if necessary to complete the work. This is exactly the relation between processors and pthreads on KSR-1.

3.6.1 Automatic Parallelization

For Fortran users, there is a preprocessor called KAP on KSR-1 that analyzes data dependence and parallelizes loops in the original source code by inserting KSR Fortran directives. It can be used stand-alone or as an option of the KSR f77 compiler. By using KAP, a user can run a sequential Fortran program on KSR-1 parallel computers without modifying the original code. The compilation process of a Fortran code using the KSR KAP is shown in Fig. **3.23**.

Figure **3.23**: Compiling process of a Fortran program prog.f using KSR
KAP

On KSR-1 parallel computers, different tasks that can be done in parallel are carried out by a team of pthreads. A program starts with one pthread and uses more pthreads as necessary to perform work in parallel. Typically, an application program creates multiple pthreads that execute simultaneously on different processors. The process is shown in Fig. 3.24. With automatic parallelization, the creation of teams of pthreads is handled automatically by the KSR run time library. The programmer can also create teams of pthreads and specify explicitly the use of a particular team of pthreads by calling library subroutines. When multiple pthreads are created, there is no guarantee that different pthreads will be assigned to different processors and

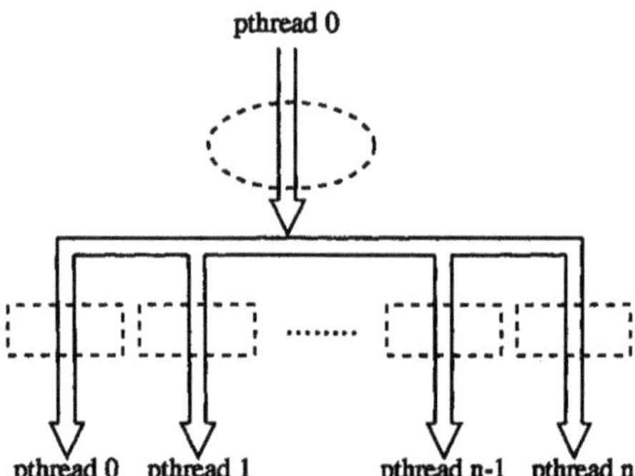

Figure 3.24: Execution of a team of pthreads. The dashed ellipse represents a sequential job processed by a single pthread, and the
dashed boxes represent parallel jobs processed by a team of
pthreads.

stay on the same processors throughout the execution, especially when there are more
pthreads than the number of processors.

If a code is parallelized by a completely automatic parallelization process, all
parallelized tasks are executed by the same default team of pthreads. Fig. 3.25
shows the progress of a program execution. The sequential parts are handled by a
single pthread, while the parallel parts are processed by multiple pthreads which all
belong to the same team of pthreads. Although in principle, the programmer can
create multiple teams of pthreads so that the tasks of parallel part 1 and parallel
part 2 in Fig. 3.25 can be executed by different teams of pthreads, it is found that
unless necessary, use of multiple teams of pthreads should be avoided. The same team
of pthreads should be used repeatedly for parallel executions to avoid the overhead
associated with the use of multiple teams of pthreads.

In the case of automatic parallelization, the number of pthreads in a team should
be set as the number of processors used, so that each pthread is assigned to a different
processor. This is very important. If the number of pthreads is larger than the number
of processors, some pthreads will migrate across different processors, which will cause
heavy data swapping. As a result, the program execution will slow down significantly.

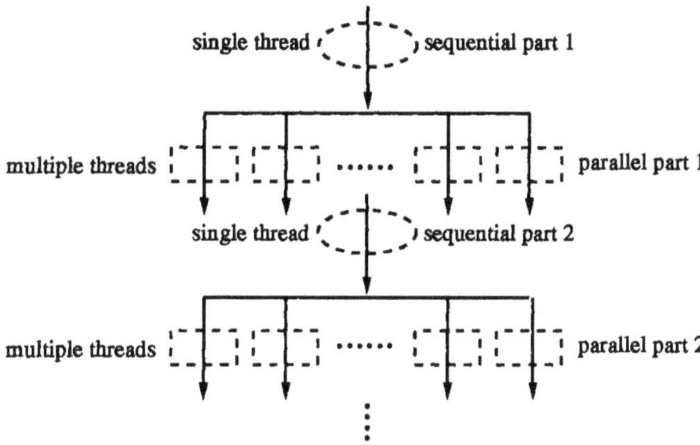

Figure 3.25: Execution flow chart of a program with both sequential and parallel parts.

3.6.2 Semi-automatic and Manual Parallelization

With KSR Fortran, a programmer can decide where and how to parallelize a sequential code. The completely automatic parallelization does not always produce satisfactory performance, particularly for large and complicated codes. The major problems involved are

- Where to parallelize. For a large code with thousands of lines, not all loops at the low level should be parallelized, especially if the loop does not encompass significant amount of work.

- How to parallelize. If a section of a code is parallelized, there are many parameters to be determined, like the number of pthreads used to execute the parallel tasks and the granularity of the parallelized tasks.

With automatic parallelization, all these are determined by the KSR compiler and the KAP preprocessor. If a programmer wants more control over where and how the code is parallelized to obtain better performance, he can use the KSR compiler directives and library subroutines to parallelize a code in a specific way. All KSR compiler directives start with "c*ksr*" at the beginning of each line, so the code is portable to other machines since all those directives will be treated as comments by other Fortran compilers.

3.6.2.1 Where to Parallelize

With semi-automatic parallelization and manual parallelization, one can use his own
judgement to determine whether a specific part of the program should be parallelized.
The parallelization of a code on KSR-1 is accomplished by using three types of parallel
compiler directives [107]:

1. Parallel Regions. The parallel region construct enables the programmer to execute parallel multiple instantiations of a single code segment. The code segment
   of a parallel region is declared by using the directives:

   ```
   c*ksr* parallel region(para1, para2, ···)
         ⋮
      code segment
         ⋮
   c*ksr* end parallel region
   ```

   where para1, para2, ···, are parameters. The code segment enclosed by the
   parallel region directive is called a parallel domain. All pthreads executing a
   parallel region execute the entire parallel domain. This, however, does not imply
   that all pthreads do exactly the same thing. As shown in the following code
   segment

   ```
   c*ksr* parallel region(numthreads=numprocs, private=(t, mynum))
          mynum = ipr_mid( )
          do i = 1, n
             if (mod(i, numprocs)·eq·mynum) then
                do j=1, n
                   t=t+A(i,j)*v(j)
                end do
                w(i) = t
             end if
          end do
   c*ksr* end parallel region
   ```

   where numprocs is the number of processors used, t and mynum are private
   variables which mean that different pthreads will have different private copies
   of these two variables, and the function ipr_mid( ) gives the pthread ID number.
   This segment of code does a matrix-vector multiplication $\mathbf{w} = A\mathbf{v}$. The number

of pthreads used in the computation is set as the number of processors in the parallel region directive. Different pthreads execute multiplications of different rows in matrix A with vector $\mathbf{v}$. The rows of matrix A are assigned to pthreads in a wrap around fashion determined by the "**if**" statement in the code. If there are totally m pthreads, then the kth pthread will be assigned row k, $k + m$, $k + 2m$, $\cdots$, in matrix A. It is clear that although the parallel domain is the same for all pthreads, different pthreads will work with different set of values of the loop index i.

2. <u>Parallel Sections</u>   Parallel sections are groups of different code segments of a program that can be run in parallel. The code segments within a parallel section are called section blocks. One pthread is created for each section block, and in general, each section block is executed on a different processor.

   Parallel sections are specified by enclosing the desired code segments within parallel section directives. Each section block in a parallel section starts with a section directive. The general semantics of a parallel section looks like

```
c*ksr* parallel sections(para1,para2,···)
c*ksr* section
       section block 1
c*ksr* section
       section block 2
          ⋮
c*ksr* section
       section block m
c*ksr* end parallel sections
```

   where there are totally m section blocks that can be executed in parallel. The private variables used in all section blocks can be specified in the parallel sections directive.

   When using the parallel sections directive, the programmer must take special care to ensure that the codes in different section blocks are independent so that they can be executed in parallel. Otherwise, using parallel sections may result in incorrect results.

3. <u>Tile families</u> The most important parallelization mechanism in KSR Fortran programming is perhaps the tile family which is used to parallelize loops. In general, most of the computationally intensive parts in a program for scientific

computations are expressed as loops. By tiling, one can transform the sequential execution of a single **do** loop to parallel execution of multiple tiles, or groups of loop iterations. The group of tiles is called a *tile family*. For example, the following loop on a two-dimensional array A

```
do i = 1, N
   do j = 1, M
      A(i,j) = ...
            :
   end do
end do
```

can be parallelized in both i and j indices if there is no data dependence. For the loop used in the discussion of parallel region on page 130, only the i index is parallelizable. The j index is not parallelizable due to data dependence. Fig. 3.26, shows that a complete iteration space $(i = 1 : N, j = 1 : M)$ is decomposed

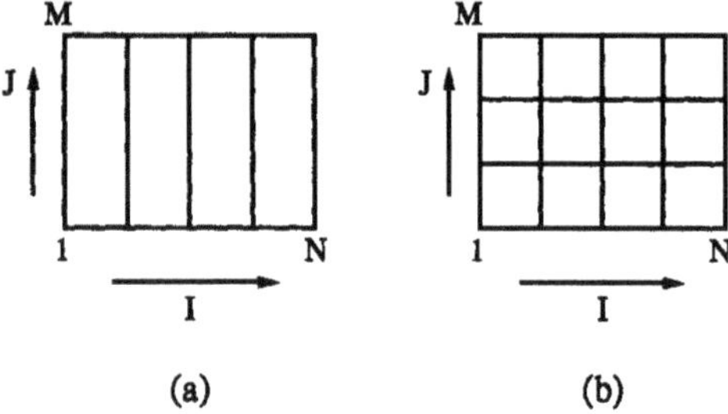

Figure 3.26: (a) The iteration space $(I = 1 : N, J = 1 : M)$ is decomposed into tiles in one index. (b) The iteration space $(I = 1 : N, J = 1 : M)$ is decomposed into tiles in two indices

into tiles which can be assigned to different processors for execution.

A tile family can be declared using the tile directive. For example, the following code

```
c*ksr* tile(i,j)
      do j=1,N
        do i=1,M
          B(i,j)=A(i,j)**2
        end do
      end do
c*ksr* end tile
```

will generate a tile family similar to that shown in Fig. 3.26(b). As in the cases for parallel region and parallel sections, many parameters can be specified in the tile directive (for example private variables) to ensure correct and efficient execution of tile families.

The three parallel constructs just discussed provide the basic means for parallel programming on KSR-1 parallel computers. In addition to that, any combination or nesting of these three parallel constructs is allowed to parallelize more complicated code segments. For example, one section block of a parallel sections construct might contain a tile family, and another section block might contain a parallel region. The basic parallel constructs are block constructs that are subject to the same rules as other Fortran block constructs.

3.6.2.2 How to Parallelize

The parallel constructs just discussed provide basic means to parallelize a code segment. Once a programmer has decided that a segment of the code should be parallelized and selected a parallel constructs to parallelize the code, there are still many parameters in that parallel construct to be determined. By specifying proper values for these parameters, a programmer can control how the code is parallelized.

For semi-automatic parallelization, a programmer does not have to specify all the parameters. The KSR-1 compiler will use the default values for the parameters not specified by the programmer. In the case of manual parallelization, all parameters are specified by the programmer. The code implementation gets more complicated as one goes from automatic parallelization to semi-automatic and manual parallelization. Correspondingly one also expects higher efficiency and better implementation as more control is added in the programming.

The following paragraphs describe some of the most important parameters in the parallel constructs that reflect the features of parallel programming on KSR-1.

1. _numthreads_. This is an important parameter that specifys the number of pthreads used to execute a particular parallel constructs. It is used in parallel

region and tile constructs. When this parameter is specified, a new team of pthreads is created. Each pthread will execute the code segment encompassed by the parallel region or tile directive once. The team of pthreads exists only for the duration of the particular parallel region or the tile construct. In other words, if there are more than one parallelized code segments in the program, as shown in Fig. 3.25, a team of pthreads will be created for each of the parallel constructs. It should be pointed out that the creation and termination of teams of pthreads incur programming overhead which could affect the execution speed.

2. _teamid_. This is an integer variable that specifies the ID of a team of pthreads created by calling the KSR intrinsic subroutine `ipr_create_team`. This subroutine creates a team of pthread with specified number of pthreads. The subroutine also returns the team ID number that can be used in parallel constructs. The following code segment creates a team of 6 pthreads and uses that team of pthreads to execute a parallel region:

```
      call ipr_create_team(6,id)
         ⋮
c*ksr* parallel region(teamid=id)
      code segment
c*ksr* end parallel region
```

This parameter can be specified in all of the three basic parallel constructs discussed before. For parallel sections directive, if the team has more pthreads than the number of section blocks, each section block will be executed by one pthread and the extra pthreads will stay idle. If there are fewer pthreads in the team than the number of section blocks, then a run-time error will occur. Since the `teamid` parameter allow the repeated use of the same team of pthreads for different parallel constructs in different code segments, it reduces the system overhead for creating, managing, and terminating many teams of pthreads for executions of different parallel constructs. Whenever possible, the same team of pthreads should be used repeatedly throughout the program execution.

3. _private_. This parameter specifies a list of variables that are not shared among different pthreads. For example the variable t used in the code segment for the discussions of the parallel region on page 130 is a private variable in different pthreads. It represents the inner product of a particular row of A with the vector **v**. Since different pthreads work on different rows of matrix A, the value of t can

not be shared among different pthreads. In the case of automatic parallelization, the KAP will analyze the code and specify private variables automatically. With semi-automatic and manual parallelization, it is the programmer's responsibility to identify and specify proper private variables to ensure that the final results are correct. An array can not be put directly in the list of private variables. If it is necessary to maintain private copies of an array for different pthreads, the "partially shared common" directive can be used to specify a private array. The following code segment shows that the array $v(1 : M)$ is first allocated in a partially shared common (PSC) block blk_1, and then the block is specified in the tile directive as private, which makes all pthreads maintain their own private copies of the array:

```
    common /blk₁/ v
  c*ksr* psc /blk₁/
  c*ksr* tile(j, teamid = id, private = (/blk₁/))
      do j=1,N
        do i=1,M
          v(i) = A(j,i)*w(i)
          ⋮
        end do
        ⋮
      end do
  c*ksr* end tile
```

Since the specification of private arrays requires the use of partially shared common blocks, the automatic parallelization is not able to parallelize code segments that require private arrays.

4. _index_list_. This variable specifies which loop indices should be parallelized (or tiled) in the tile directive. As shown in Fig. 3.26, a two-index do-loop can be parallelized in index i (Fig. 3.26(a)), or index j, or both index i and j (Fig. 3.26(b)). Which index should be parallelized depends on the analysis of data dependence, load balance, and memory access patterns.

5. _tilesize_. This parameter is used in the tile directive to specify the size along each tile index. If this parameter is used, a tile size for each tile index must be used. The following code segment shows a tile size of 6 and 12 in the i and j indices respectively:

```
c*ksr* tile(i, j, tilesize=(i:6, j:12))
      do j = 1, N
        do i = 1, M
          ⋮
        end do
      end do
c*ksr* end tile
```

If this parameter is not specified, then the tile size is determined by the KSR run-time subroutines with two objectives: maximizing parallelism and making the best use of the memory system.

Generally speaking, a tile family with a smaller tile size has a higher degree of parallelism than a tile family with a larger tile size. But it incurs more parallel overhead. In the case that not all processors are dedicated to the running program, a tile family with a small tile size can utilize the processors more efficiently with better load balance than does the tile families with large tile sizes. If all processors are used exclusively for running a particular program and all processors are available for executing a tile family, it is more efficient to use a tile size as large as possible to create just as many tiles as the number of available processors [203].

6. *strategy*. This is a parameter used only in the parallel construct of tile family. It determines how the iteration (tile) space is divided and assigned to different processors. There are four different tiling strategies for different situations:

   (a) *Slice Strategy*. The slice strategy evenly divides the iteration space into n tiles, where n is the number of pthreads used to execute the tile family. If the iteration space can not be divided evenly into n tiles, the tile size will be rounded up to a larger whole number. For example, if the loop index to be tiled runs from 1 to 100 and three pthreads are used, then the slice strategy will produce three tiles with a size of 34, except that the last tile will actually contain only the left-over 32 iterations from the loop.

   The slice strategy minimizes tile-creation overhead and the possibility of data contention at tile boundaries, since there are only minimum number of tiles (to keep all the pthreads busy) created.

   It appears that specifying the tilesize parameter with a size as large as possible (but still keeping all pthreads busy) will also produce the same tile family as does the slice strategy. For example, with a loop index going from 1 to 100, specifying a tilesize of 34 should also produce three tiles with sizes of 34, 34, and 32 respectively. However, numerical experiments show that using the tilesize parameter to generate large tiles produces more efficient code than does the slice strategy [203].

(b) *Modulo Strategy.* The modulo strategy typically creates many more tiles (thus with small sizes) than the number of pthreads in the team. The assignment of tiles to pthreads is done based on the operation $mod(i, n)$ where n is the number of pthreads and i is the tile ID number in the tile family. Therefore, with ten tiles and three pthreads, pthread 0 will take tiles 0, 3, 6, 9, pthread 1 will take tiles 1, 4, 7, and pthread 2 takes tiles 2, 5, and 8. If the same arrays are used in several different tile families, using the modulo strategy produces the same pthread-to-tile assignment, which reduces possible data exchange between different processors.

(c) *Wavefront Strategy.* This strategy is used to parallelize loops that have certain dependences between different iterations. For example, the loop

```
do j = 2, N-1
   do i = 2, N-1
      a(i,j) = a(i-1,j)+a(i,j-1)
   end do
end do
```

has the data dependence similar to that of solving the Laplace equation using the Gauss-Seidel iterative algorithm. It has been discussed in Section 2.2.3 that all loops with $i + j = constant$ can be executed in parallel. The computation progresses like a wavefront shown in Fig. 2.6(b). The wavefront strategy parallelizes tiles of a tile family in a similar fashion. All tiles on a diagonal of the iteration space will be executed in parallel.

(d) *Grab Strategy.* This strategy assigns tiles to pthreads dynamically based on the availability of pthreads and processors. Tiles are distributed to pthreads on a first-come, first-served basis. Therefore, it is suitable for execution of a tile family with unbalanced load caused, for example, by the non-exclusive use of processors.

Chapter

4

Applications

In this Chapter, we will discuss parallelization and implementation of algorithms for solving two problems on Intel iPSC/860 hypercubes and Kendall Square KSR-1 parallel computers. The first problem is the numerical solution of Poisson's equation, and the second problem is the simulation of multiphase flow in porous media.

4.1 Numerical Solution of Poisson's Equation

In this section, we discuss implementations of numerical algorithms for solving the Poisson's equation on parallel computers. Although the equation is simple, it is widely used in scientific and engineering applications, mostly static field problems. The solution algorithms used here are the Jacobi and Gauss-Seidel relaxations. The discretized equation has the form

$$\frac{u_{i,j-1} + u_{i-1,j} - 4u_{i,j} + u_{i+1,j} + u_{i,j+1}}{h^2} = f_{i,j} \qquad (4.1)$$

$$i, j = 1, 2, \cdots n,$$

for a two-dimensional uniform grid with grid spacing h. The Jacobi iteration for Eq. (4.1) can be written pointwisely as

$$u_{i,j}^{(k+1)} = \frac{1}{4}(u_{i,j+1}^{(k)} + u_{i-1,j}^{(k)} + u_{i+1,j}^{(k)} + u_{i,j+1}^{(k)} - h^2 f_{i,j}) \qquad (4.2)$$

$$i, j = 1, 2, \cdots n, \quad k = 0, 1, \cdots,$$

which is perfectly parallelizable.

The original two-dimensional domain is decomposed into subdomains which are assigned to different processors. Each processor updates the values at the grid points in its own subdomain. In the end of each iteration, a processor needs to exchange the values at the grid points on the boundary of its own subdomain with the processors holding the adjacent subdomains.

In the numerical experiment, the P processors used in the computations are arranged as a $P = P_x \times P_y$ grid to accommodate different domain decomposition schemes. When $P_y = 1$, it corresponds to a strip decomposition shown in Fig. 4.1(a), while $P_x = 1$ corresponds to a strip decomposition shown in Fig. 4.1(b). The rea-

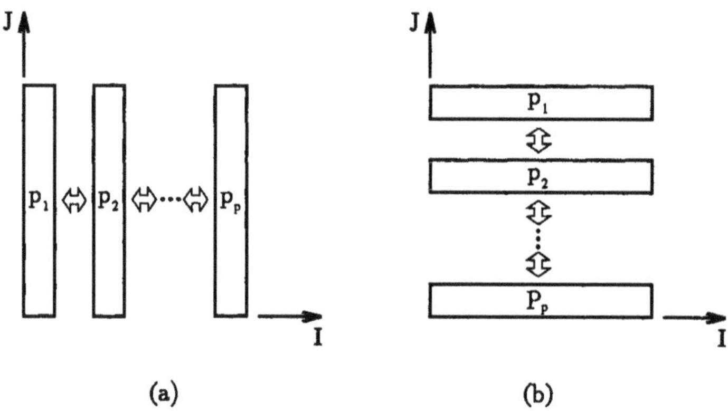

Figure 4.1: Two strip domain decomposition schemes. (a) $P_y = 1$ in $P = P_x \times P_y$, (b) $P_x = 1$ in $P = P_x \times P_y$.

son for using these two actually equivalent decompositions is to see the influence of data storage on inter-processor communications. If the data in all subdomains of Fig. 4.1(a) and Fig. 4.1(b) are stored in arrays in the same way, then the data to be transferred in each iteration are columns of arrays in Fig. 4.1(a) and rows of arrays in Fig. 4.1(b). With Fortran programming language, elements in an array are mapped by columns. Therefore, a column of an array can be exchanged directly in the inter-processor communication process on the iPSC/860 hypercube, while a row of an array can not be exchanged directly. It must be packed into a one-dimensional array before it can be exchanged. This issue was discussed in detail in Section 3.3.1.

Table 4.1 shows the timing results for solving the Poisson's equation on a two-dimensional uniform grid using the Jacobi iteration. The convergence criterion was

Table 4.1: Timing results (seconds) for solving
Poisson's Equation on iPSC/860
using the Jacobi iterations

$n \times n$ / $p_x \times p_y$	32×32	64×64	128×128
1×1	7.20	85.4	949
2×1	4.56	45.7	485
1×2	4.72	46.5	490
4×1	3.52	27.2	258
1×4	3.82	28.6	266
2×2	2.73	27.2	256
8×1	2.95	17.5	144
4×2	2.16	17.5	144
2×4	2.26	18.2	149
1×8	3.25	19.1	152

set as 10^{-4} for the maximum residual norm.

The results are presented for $P = 1$, 2, 4, and 8 processors. For the cases of $P = 4$ and $P = 8$, the processors are arranged as either a chain or a two-dimensional grid. From the table, we can see clearly that $P_y = 1$ is a much better decomposition (Fig. 4.1(a)) than the case of $P_x = 1$, which is not surprising from the early discussions. This suggests that whenever possible, the data to be exchanged should be stored as columns of matrices, rather than as rows of matrices. It is also clear from the table that when more processors are used in the computation, two-dimensional domain decomposition can be more efficient for inter-processor communications, since it generates subdomains with smaller circumference/area ratio. For example, in the case of four processors, Table 4.1 shows that the case $P_x \times P_y = 2 \times 2$, which corresponds to decomposing the original domain into 4 patchs, gives better timing results than the cases of $P_x \times P_y = 4 \times 1$ or $P_x \times P_y = 1 \times 4$. As shown in Fig. 4.2, the message length in each communication (which is proportional to the length of one side of a subdomain) for the patch decomposition is only half of that for the strip decompositions. Although in the patch decomposition, each of the four processors communicates with two processors, while in the strip decomposition only two of the four processors communicate with two processors and the other two communicate with only one processor, the overall time spent on communication is shorter for the patch decomposition because of the better circumference/area ratio. If the domain is not a square, then it is not necessary to use the two-dimensional patch decomposi-

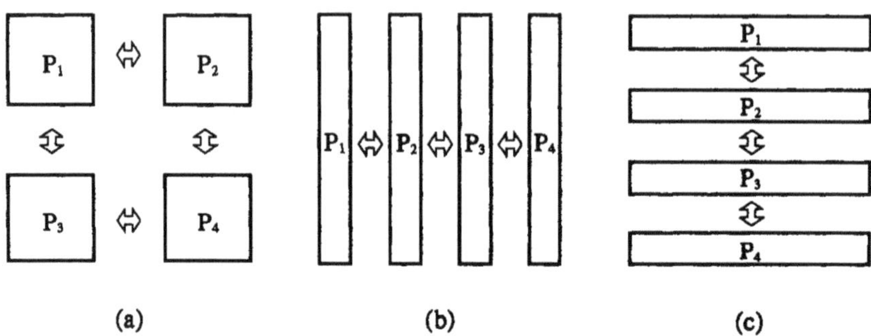

Figure 4.2: Three different domain decompositions. (a) 2×2 patch decomposition. (b) 4×1 strip decomposition. (c) 1×4 strip decomposition.

tion. A one-dimensional strip decomposition may produce subdomains with shorter message length than that generated by the two-dimensional patch decomposition, as shown by Fig. 4.3. The original domain in Fig. 4.3(a) is a narrow rectangle, therefore the one-dimensional strip decomposition in Fig. 4.3(b) generates subdomains with better circumference/area ratio than the two-dimensional patch decomposition in Fig. 4.3(c).

With eight processors, we see from Table 4.1 that the case of $P_x \times P_y = 8 \times 1$ is as good as or better than the other three cases. That is because with 4×2 and 2×4 decompositions, some processors need to communicate with three processors, while in the strip decomposition, a processor communicates with at most two processors.

Fig. 4.4 shows the speedup curves for the timing results given in Table 4.1. The speedup is calculated as the ratio of the sequential execution time to the parallel execution time. For a given number of processors, the best execution time among those obtained using different decompositions is used as the parallel execution time. In Fig. 4.4, we can see that the larger the problem, the better the speedup. This is true in general for many numerical algorithms for solving PDEs. In order to utilize the increasing number of processors, one must increase the problem size.

Fig. 4.5 shows the speedup curves for solving the same equation using the Gauss-Seidel iterations. In Section 2.2.3.2, the red-black ordering was discussed for parallelizing the Gauss-Seidel type relaxations. Here we used a different strategy to implement the algorithm. The implementation is based on decoupling the data de-

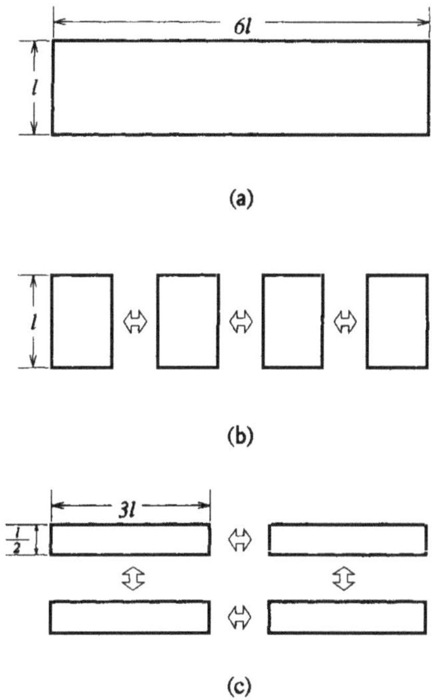

Figure 4.3: (a) the original rectangular domain. (b) one-dimensional strip decomposition produces subdomains with shorter message length. (c) two-dimensional patch decomposition produces subdomains with longer message length.

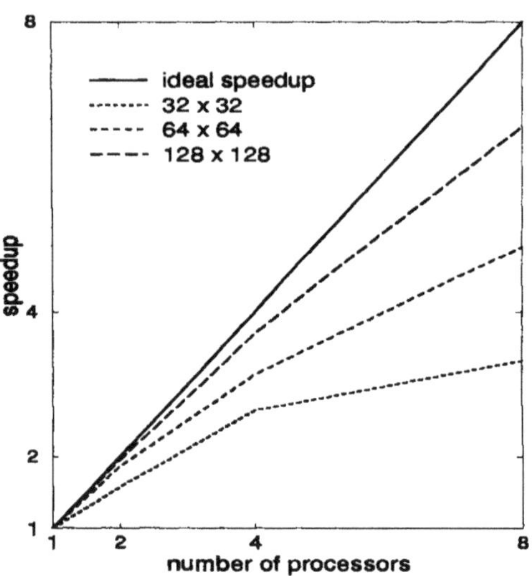

Figure 4.4: Speedup curves for the timing results given in Table 4.1

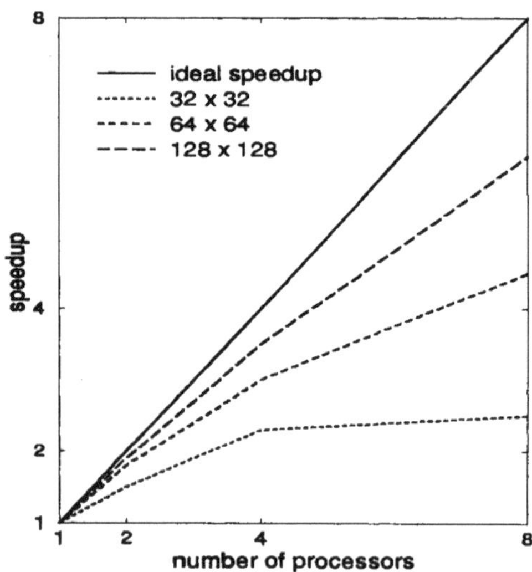

Figure 4.5: Speedup curves obtained using the Gauss-Seidel algorithm.

pendence of the Gauss-Seidel iterations between subdomains while retaining the data dependence inside each subdomain. As a result of this decoupling, the number of iterations needed to meet the same convergence criterion increases as the number of processors increases. In the extreme case that each processor has only one grid point, the algorithm degenerates to the Jacobi iterations. In the case when there are many grid points in a subdomain, there is only very limited degradation in the convergence rate. For example, with a 128×128 grid which is decomposed as 8 strips, there is only five percent increase in the number of iterations with 8 processors as compared to that with only one processor.

4.2 Numerical Simulations of Multi-phase Flow

Numerical simulations of multiphase flow are widely used in many engineering problems [196, 203]. A typical example is reservoir simulations [49, 50, 186].

Fig. 4.6 shows a reservoir model with production and injection wells. For any

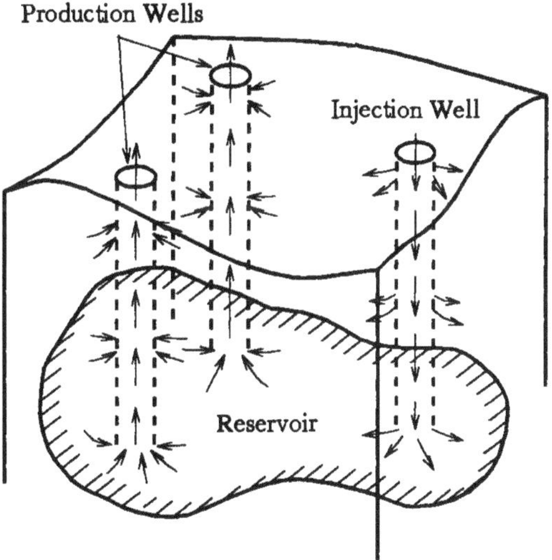

Figure 4.6: A reservoir with injection and production wells

production well in an oil reservoir, the production rate (the rate at which oil is produced) gradually decreases due to the declining reservoir pressure, and eventually the well becomes depleted. The oil production process in this period is called the primary oil recovery process in which oil is produced by natural reservoir pressure. Scientific study shows that only a small portion (15–25%) of the oil in a reservoir can be recovered this way. To maximize oil production, various secondary and enhanced oil recovery processes have to be used to force the oil in a reservoir to flow out of production wells. If properly designed and implemented, the use of these processes can increase the oil production by at least 10–15% in addition to the oil produced in the primary recovery process.

The central part in various enhanced oil recovery processes is the injection of water or other chemicals into a reservoir to build up the pressure gradient or to increase the mobility of the oil. The moving fluid front between the injected fluid and the oil should move in such a way as to drive the oil out of production wells. The stability of this moving front is crucial to the success of the enhanced oil recovery process. If the front becomes unstable, it will no longer displace oil toward the production wells. Fig. 4.7 shows the "fingering phenomenon" [65] in a two-dimensional reservoir. The

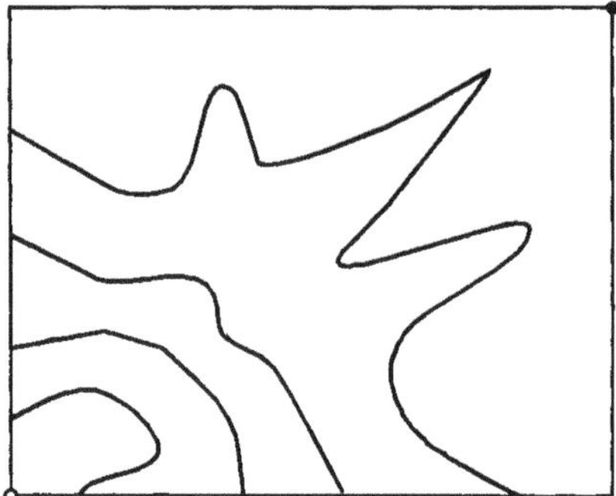

Figure 4.7: The fingering phenomenon. The long narrow "fingers" represent the unstable injected fluid front bypassing the oil. The injection well is located in the lower left corner and the production well is located in the upper right corner.

injection well is located in the lower left corner and the production well in the upper right corner of the figure. The contour lines represent the front between the injected fluid and the oil at different times. The injected fluid is supposed to displace oil toward the production well. When the front becomes unstable, however, the injected fluid can move much faster than the oil and bypass the oil, as indicated by the contour lines in the figure. This is called the "fingering phenomenon". As a result, what comes out of the production well might be the injected fluid, rather than oil. To prevent this from happening and to make enhanced oil recovery processes productive, we must be able to monitor and predict the fluid front movement during the injection process.

Since reservoirs are usually deep under the ground, the movement of the fluid front between the injected fluid and the oil is not directly observable. Therefore, mathematical models are very important for keeping track of the changes in reservoir pressure and the movement of the fluid front.

In the commonly used black-oil model [148], a three-dimensional two-phase (oil and water) reservoir can be described by:

$$\nabla \cdot [\lambda_o K(\nabla P_o - \rho_o \mathbf{g})] = \frac{\partial}{\partial t} \left(\frac{\Phi S_o}{B_o} \right) + q_o,$$

$$\nabla \cdot [\lambda_w K(\nabla P_w - \rho_w \mathbf{g})] = \frac{\partial}{\partial t} \left(\frac{\Phi S_w}{B_w} \right) + q_w, \tag{4.3}$$

$$t > 0, \quad \underline{x} = (x, y, z) \in \Omega,$$

$$S_o + S_w = 1, \qquad P_{cow} = P_o - P_w = f(S_w), \tag{4.4}$$

with the initial conditions

$$P_l(x, y, z, 0) = P_l^0(x, y, z),$$

$$l = o, w, \quad \underline{x} \in \Omega,$$

and the boundary conditions

$$\frac{\partial P_l}{\partial \underline{n}} = 0, \quad l = o, w, \quad \underline{x} \in \partial\Omega, \ t > 0, \tag{4.5}$$

where the transmissibilities are defined by

$$\lambda_l = \frac{k_{rl}}{\mu_l B_l}.$$

The subscripts o and w refer to the phases of oil and water respectively, all other symbols have the following meanings:

P_l: phase pressure, $l = o, w$

S_l: phase saturation, $l = o, w$

K: permeability, function of $\underline{x}$

k_{rl}:   relative permeability, function of saturation S_l

B_l: Format Volume Factor, function of P_l

μ_l: viscosity of phase l, assumed as a constant

g: gravitational acceleration vector

ρ_l: fluid density

Φ: rock porosity, assumed as a constant

q_l: volumetric production rate per unit reservoir volume (negative if injection)

$f(S_w)$: functional relation between capillary pressure and water saturation

Eq. (4.3) was derived from the conservation laws, the equations of states and the Darcy's law. The equations are nonlinear since k_{rl}'s are functions of saturation S_l, and B_l's are functions of pressure P_l. There are totally four unknowns, P_o, P_w, S_o and S_w in two partial differential equations. With two additional functional relations (4.4), the Eq. (4.3) can be solved for pressure and saturation distributions. Here the so called SS (simultaneous solution) method [13] is used to solve Eqs. (4.3)–(4.5). The key step is to use the functional relations (4.4) to eliminate the saturations in Eq. (4.3) and thus establish a new system that contains pressures P_o and P_w only:

$$\nabla \cdot [\lambda_o K(\nabla P_o - \rho_o g)] = \sum_{l=o,w} C_{o,l}\frac{\partial P_l}{\partial t} + q_o$$

$$\nabla \cdot [\lambda_{w} K(\nabla P_w - \rho_w g)] = \sum_{l=o,w} C_{w,l}\frac{\partial P_l}{\partial t} + q_w \tag{4.6}$$

The derivation of Eq. (4.6) is based on the assumption that there is a unique inverse function f^{-1} so that $S_w = f^{-1}(P_{cow})$. In most practical situations, this relation is given in the form of experimental data and tables. All the coefficients C_{ij}'s can be obtained by using the chain rule for differentiation. Take the first equation in Eq. (4.3) as an example, the right hand side can be written as

$$
\begin{aligned}
\frac{\partial}{\partial t}(\frac{\Phi S_o}{B_o}) + q_o &= \Phi S_o\frac{\partial}{\partial t}(\frac{1}{B_o}) + \frac{\Phi}{B_o}\frac{\partial S_o}{\partial t} + q_o \\
&= \Phi S_o\frac{\partial}{\partial P_o}(\frac{1}{B_o})\frac{\partial P_o}{\partial t} + \frac{\Phi}{B_o}\frac{\partial S_o}{\partial P_{cow}}\frac{\partial P_{cow}}{\partial t} + q_o \\
&= \Phi S_o\frac{\partial}{\partial P_o}(\frac{1}{B_o})\frac{\partial P_o}{\partial t} + \frac{\Phi}{B_o}\frac{\partial(1-S_w)}{\partial P_{cow}}(\frac{\partial P_o}{\partial t} - \frac{\partial P_w}{\partial t}) + q_o \\
&= \left[\Phi S_o\frac{\partial}{\partial P_o}(\frac{1}{B_o}) - \frac{\Phi}{B_o}\frac{\partial S_w}{\partial P_{cow}}\right]\frac{\partial P_o}{\partial t} + \frac{\Phi}{B_o}\frac{\partial S_w}{\partial P_{cow}} \cdot \frac{\partial P_w}{\partial t} + q_o,
\end{aligned}
$$

so that

$$C_{o,o} = \Phi S_o\frac{\partial}{\partial P_o}(\frac{1}{B_o}) - \frac{\Phi}{B_o}\frac{\partial S_w}{\partial P_{cow}},$$

$$C_{o,w} \; = \; \frac{\Phi}{B_o}\frac{\partial S_w}{\partial P_{cow}}.$$

Applying the same procedure to the second equation in Eq. (4.3), we get

$$C_{w,o} \; = \; \frac{\Phi}{B_w}\frac{\partial S_w}{\partial P_{cow}},$$

$$C_{w,w} \; = \; \Phi S_w \frac{\partial}{\partial P_w}\left(\frac{1}{B_w}\right) - \frac{\Phi}{B_w}\frac{\partial S_w}{\partial P_{cow}}.$$

4.3 Discretization and Parallelization

Finite difference scheme is used to discretize the spatial domain and the equations. The spatial and temporal derivatives are discretized by the 7-point central differencing and the implicit backward differencing schemes respectively. The domains discussed here are all rectangular with uniform grid for simplicity. There are four unknowns associated with each grid point. As described previously, only P_o and P_w are solved simultaneously from Eq. (4.6). The other two unknowns S_o and S_w can be solved from the functional relation (4.4).

After discretization, Eq. (4.6) becomes a nonlinear algebraic equation system

$$(\mathbf{T} - \mathbf{D})_{m+1}\mathbf{p}_{m+1} = -\mathbf{D}\mathbf{p}_m + \mathbf{q}_{m+1} \tag{4.7}$$

where

$$\mathbf{p} = \{p_{o1}, p_{w1}, p_{o2}, p_{w2}, \cdots\cdots, p_{oN}, p_{wN}\}^T.$$

Note that there are two pressure values P_o and P_w associated with each grid point, N is the total number of grid points. T is a $2N \times 2N$ 7 block diagonal matrix, D is a $2N \times 2N$ block diagonal matrix, and q is a $2N \times 1$ production vector. The subscript m indicates time steps. Since there are two equations at each grid point, all the entries in matrices T and D of Eq. (4.7) are 2×2 block matrices. For simplicity, we will still use the word "entry" or "element" when discussing matrices T and D, with the understanding that the word really means a 2×2 block.

Eq. (4.7) can be solved by the Newton's method or the linearization method discussed in [13]. For simplicity, we omit the indices representing different time steps and different iterations at each time step respectively, and write the linearized version of Eq. (4.7) at each time step as:

$$\mathbf{M}\mathbf{p} = \mathbf{f}$$

where the structure of M depends on the ordering of the grid points, the specific discretization scheme, and the linearization method. When the second order central space finite differencing, the first order implicit time differencing, and the linearization method of [13] are used with the lexicographic grid ordering, we will have:

$$\begin{bmatrix} A_1 & B_1 & & \\ C_1 & \ddots & \ddots & \\ & \ddots & \ddots & B_{n-1} \\ & & C_{n-1} & A_n \end{bmatrix} \begin{bmatrix} p_1 \\ \vdots \\ \vdots \\ p_n \end{bmatrix} = \begin{bmatrix} f_1 \\ \vdots \\ \vdots \\ f_n \end{bmatrix} \tag{4.8}$$

where n is the number of vertical grid lines (or horizontal lines), p_i represents the vector of unknowns of the grid points on the ith vertical (or horizontal) grid line, and all submatrices are themselves sparse matrices. For two-dimensional domains, A_i is a block tri-diagonal matrix, and C_i and B_i are block diagonal matrices. Each block in A_i, B_i and C_i is a 2×2 submatrix corresponding to two unknowns at each point. By solving this equation at different time steps, we can obtain values of pressures. The saturations can be computed by using Eq. (4.4).

For large models, Eq. (4.8) is usually solved by iterative methods. Here we discuss the use of Gauss-Seidel and SOR type relaxation schemes which can be accelerated by the multigrid method. The major part is to parallelize the smoothing process using Gauss-Seidel and SOR type relaxations.

It has been discussed in Section 2.2.3.2 that the Gauss-Seidel type relaxation process can be parallelized by using the red-black ordering as shown in Fig. 2.7. For better convergence and larger granularity in parallel processing, the line or block Gauss-Seidel (SOR) relaxation with red-black ordering can be used in a similar fashion like point relaxation on parallel computers. Fig. 4.8 shows the line relaxation with red-black ordering on a two-dimensional domain. The high-lighted lines are red lines and all grid points on these lines are ordered first, followed by the grid points on the thin black lines. Fig. 4.9 shows the block red-black ordering which is also called checkboard ordering. All points in the high-lighted red blocks are ordered first, followed by the points in the black blocks. The corresponding relaxation schemes should be the block relaxations. The matrix structure for the reordered system is in the form of

$$\begin{bmatrix} D_r & F_b \\ G_r & D_b \end{bmatrix} \left\{ \begin{array}{c} p_r \\ p_b \end{array} \right\} = \left\{ \begin{array}{c} f_r \\ f_b \end{array} \right\}$$

where D_r and D_b are block diagonal matrices. For the line red-black ordering, each block in D_r and D_b is a tri-diagonal matrix corresponding to a line, while for block red-black ordering, each block is a penta-diagonal matrix corresponding to a patch in Fig. 4.9. For matrices F_b and G_r, with the line red-black ordering, they are block bi-diagonal matrices representing connections of a line with two adjacent lines of

Figure 4.8: Line relaxation with red-black ordering, r_i and b_i represent red and black lines, respectively.

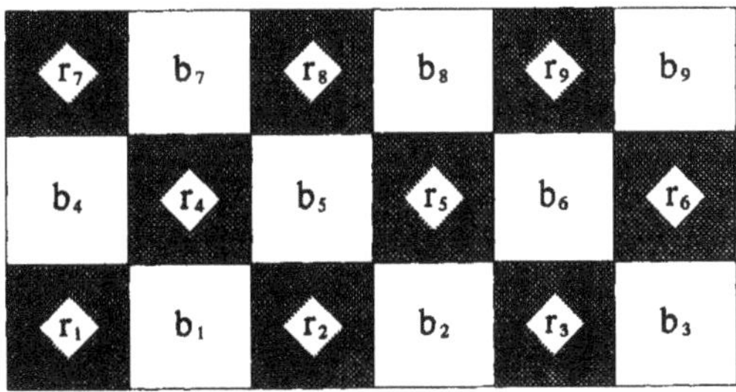

Figure 4.9: Block relaxation with red-black ordering, r_i and b_i represent red and black blocks, respectively.

different colors. In the case of block red-black ordering, matrices F_b and G_r are block 4-diagonal matrices representing the connections of a block with four adjacent blocks of different colors.

We now develop parallel reservoir simulation algorithms. With line red-black ordering and a two-dimensional domain, we can rewrite Eq. (4.8) as:

$$
\begin{bmatrix}
A_{r_1} & & & & B_{b_1} & & & \\
& \ddots & & & C_{b_1} & \ddots & & \\
& & \ddots & & & \ddots & \ddots & \\
& & & A_{r_m} & & & C_{b_{m-1}} & B_{b_m} \\
D_{r_1} & E_{r_2} & & A_{b_1} & & & & \\
& \ddots & \ddots & & & \ddots & & \\
& & \ddots & E_{r_m} & & & \ddots & \\
& & & D_{r_m} & & & & A_{b_m}
\end{bmatrix}
\begin{bmatrix}
p_{r_1} \\ \vdots \\ \vdots \\ p_{r_m} \\ p_{b_1} \\ \vdots \\ \vdots \\ p_{b_m}
\end{bmatrix}
=
\begin{bmatrix}
f_{r_1} \\ \vdots \\ \vdots \\ f_{r_m} \\ f_{b_1} \\ \vdots \\ \vdots \\ f_{b_m}
\end{bmatrix}
\tag{4.9}
$$

where subscripts r_i and b_i refer to the unknowns on red grid line r_i and black grid line b_i respectively, and we have assumed that the number of grid lines n is an even number $2m$. If the line SOR method is used to solve Eq. (4.9), we have:

$$
\begin{aligned}
p_{r_1}^{k+1} &= \omega p_{r_1}^k + (1-\omega)A_{r_1}^{-1}(f_{r_1} - B_{b_1}p_{b_1}^k) \\
p_{r_i}^{k+1} &= \omega p_{r_i}^k + (1-\omega)A_{r_i}^{-1}(f_{r_i} - B_{b_i}p_{b_i}^k) - C_{b_{i-1}}p_{b_{i-1}}^k
\end{aligned}
\tag{4.10}
$$

$$
i = 2,\, 3,\, \cdots,\, m, \quad k = 0,\, 1,\, \cdots
$$

and

$$
\begin{aligned}
p_{b_m}^{k+1} &= \omega p_{b_m}^k + (1-\omega)A_{b_m}^{-1}(f_{b_m} - D_{r_m}p_{r_m}^{k+1}) \\
p_{b_i}^{k+1} &= \omega p_{b_i}^k + (1-\omega)A_{b_i}^{-1}(f_{b_i} - D_{r_i}p_{r_i}^{k+1}) - E_{r_{i+1}}p_{r_{i+1}}^{k+1}
\end{aligned}
\tag{4.11}
$$

$$
i = 1,\, 2,\, \cdots m-1, \quad k = 0,\, 1,\, \cdots.
$$

The reservoir simulation algorithm can then be described at a very coarse grain level as:

Algorithm 4.1
Reservoir Simulation Algorithm

Initialization, set $p_{r_i}^0$ and $p_{b_i}^0$
do at each time step $k+1$
    1. **for** $i = 1,\ 2m$
        form all block equations represented
        by Eqs. (4.10) and (4.11)
    end for
    do until converge
        2. **for** $i = 1,\ m$
            compute $p_{r_i}^{k+1}$ by solving Eq. (4.10)
        end for
        3. **for** $i = 1,\ m$
            compute $p_{b_i}^{k+1}$ by solving Eq. (4.11)
        end for
    end do
end do

Note that in step 1, all block equations can be formed simultaneously. In step 2, all the red unknowns $p_{r_i}^{k+1}$ can be solved in parallel in each SOR iteration from Eq. (4.10). After that, all the black unknowns $p_{b_i}^{k+1}$ can also be solved in parallel from Eq. (4.11). Suppose m is divisible by the number of processors p, then each processor can take m/p red blocks and m/p black blocks. In each iteration, a processor first solves its share of $p_{r_i}^{k+1}$'s, and then solves its share of $p_{b_i}^{k+1}$'s.

There are three places where inter-processor communication is needed to coordinate the computations in each iteration:

1. Verification of the convergence criterion. The most commonly used criterion is to specify a given tolerance for the norm of the residual vector of Eq. (4.9). Each component in the residual vector represents the deviation by which the approximate solution fails to satisfy the corresponding equation. If the infinity norm $||\cdot||_\infty$ is used to calculate the residual norm, a processor has to first calculate the residual components associated with the equations held by that processor, and then selects the components with the maximum absolute value e_i. These local maximum residual components e_i on different processors must be compared to each other so that the largest component can be selected. The comparison and

selection process requires communication between different processors.

2. After step 2 in algorithm 4.1, the value of $p_{r_i}^{k+1}$ on the boundary of a subdomain must be passed to the processor holding the adjacent subdomain for the calculations in step 3. As shown in Eq. (4.11), the computation of $p_{b_i}^{k+1}$ requires $p_{r_{i+1}}^{k+1}$. For the domain given in Fig. 4.8, if there are four processors and each processor is assigned 4 consecutive lines (two red and two black lines), then processors 1, 2, and 3 must pass $p_{r_3}^{k+1}$, $p_{r_5}^{k+1}$, and $p_{r_7}^{k+1}$ to processors 0, 1, and 2 respectively, as shown in Fig. 4.10

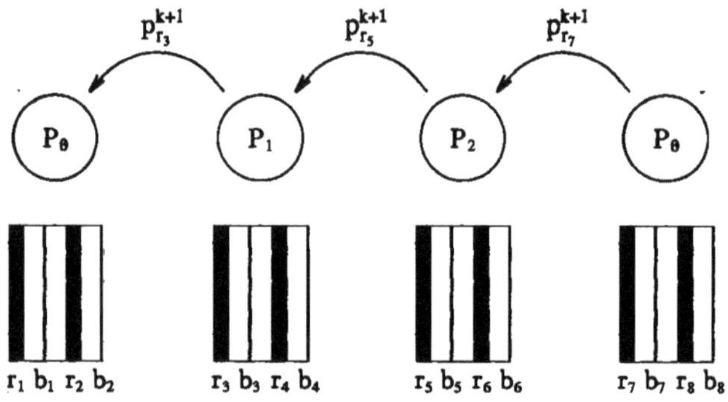

Figure 4.10: Each processor is assigned two red and two black lines, processor P_i $(i = 1, 2, 3)$ sends a message to P_{i-1}.

3. After step 3, the value of $p_{b_i}^{k+1}$ on the boundary of a subdomain must be passed to the processor holding the adjacent subdomain for the computation of step 2 in the next iteration. As shown in Eq. (4.10), the computation of $p_{r_i}^{k+1}$ requires $p_{b_i}^{k}$ which is the value calculated from step 2 of the previous iteration. For the same domain as shown in Fig. 4.8 and four processors, the communication pattern is shown in Fig. 4.11, in which processors 0, 1, and 2 pass $p_{b_2}^{k+1}$, $p_{b_4}^{k+1}$, and $p_{b_6}^{k+1}$ to processors 1, 2, and 3 respectively. Note that the $p_{b_{i-1}}^{k+1}$'s passed in Fig. 4.11 are used as $p_{b_{i-1}}^{k}$'s in the next iteration.

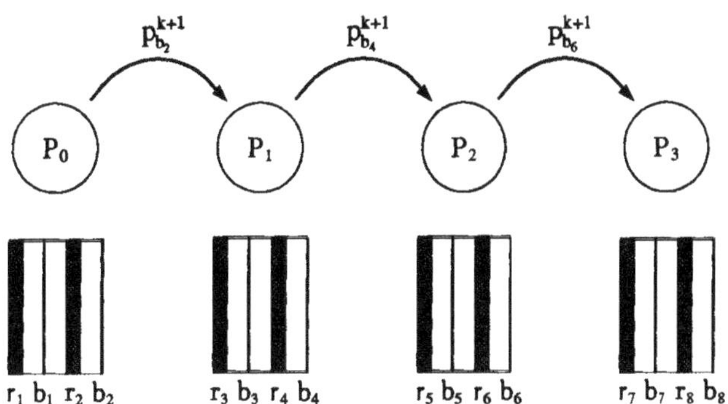

Figure 4.11: Each of the four processors is assigned two red and two black lines, processor P_i $(i = 0, 1, 2)$ sends a message to P_{i+1}.

4.4 Numerical Experiments

This section presents timing results of solving the porous medium flow model on Intel iPSC/860 hypercubes and Kendall Square KSR-1 parallel computers. For each architecture, a few different implementations of the same parallel algorithm will be discussed and compared to demonstrate how the algorithm performance can be improved in a progressive manner. On iPSC/860 hypercubes, we will concentrate on how to minimize the time spent on inter-processor communications. On KSR-1 parallel computers, our focus will be on the use of semi-automatic parallelization strategies for better performance.

4.4.1 Results on iPSC/860 Hypercubes

The parallel algorithm for porous medium flow simulation discussed in the previous section was implemented on iPSC/860 hypercubes. A two-dimensional $n \times n$ reservoir model was used to carry out numerical experiments. The timing results reported here are the times, in seconds, required to advance the solution over one time step. Three different implementations will be discussed to demonstrate the efficiency improvement by reducing the time for verification of the convergence criterion and overlapping communications with computations.

Case 1. Assume that the model size is $n \times n$ with $n = 2m$, and m is divisible by the number of processors P used in the computation. With line red-black ordering, each processor is assigned $l = m/P$ red and black lines. Fig. 4.12, shows the

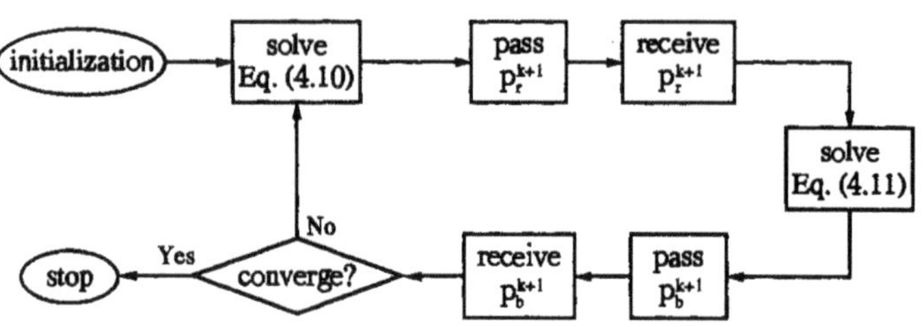

Figure 4.12: Flow chart of implementation Case 1 of the reservoir simulation algorithm on iPSC/860.

flow chart of the program execution on a processor which is not at one of the two ends of the one-dimensional chain as shown in Fig. 4.10 and Fig. 4.11. For the processors at the two ends, they either receive a message or send a message, but not both. For verification of the convergence criterion, all processors send their calculated residual components e_i to the host machine, and the host checks whether the convergence criterion has been satisfied. The number of communications needed to verify the convergence criterion is proportional to the number of processors P used in the computation.

The verification scheme on the host machine can be described as:

```
              Verification Algorithm 1

do i = 1, num_procs
      receive e_i from processor P_i
end do
e = max{e_1, e_2, ..., e_num_procs}
if (e < tolerance) then
      notify all processors to stop
end if
```

Note that in the above scheme, inter-processor communication is needed not only for sending e_i's to the host, but also for notifying all processors to continue or stop.

Table 4.2 shows the timing results on iPSC/860 hypercubes with this imple-

Table 4.2: Timing results of simulation Case 1 on iPSC/860.
P is the number of processors,
n is the grid size.

n \ P	1	2	4	8	16	32
128×128	22.1	11.3	6.0	3.5	2.8	3.1
128×384	65.2	32.9	16.8	8.9	5.47	4.6

mentation. In the table, P is the number of processors used, n is the size of the two-dimensional grid, and the results are shown in seconds. We see that as the number of processors increases, the execution time is significantly reduced. Fig. 4.13 gives the speedup curves for the timing results in Table 4.2. It is clear

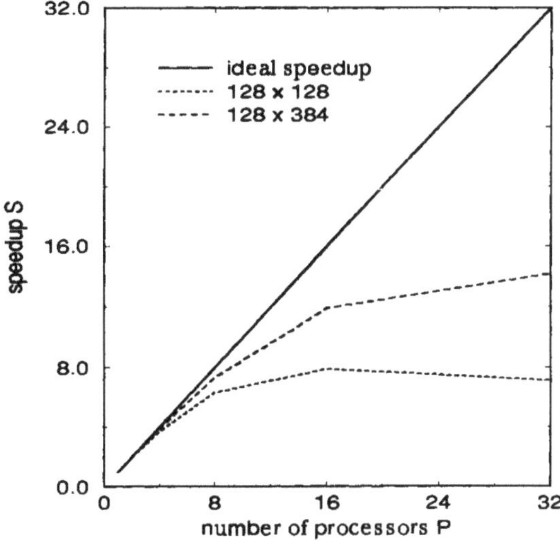

Figure 4.13: Speedup curves for simulation Case 1 on iPSC/860.

that the speedup curves deviates from the ideal speedup curve significantly as the number of processors increases. This is mainly due to the fact that as the number of processors increases, there is less computation on each processor, but more communication overhead for passing $p_{r_i}^{k+1}$ and $p_{b_i}^{k+1}$ and for verifying the convergence criterion. The situation gets better as the model size increases.

Case 2. In this case, a more efficient communication scheme is used for verification of the convergence criterion. Instead of making all processors send residual components e_i's to the host and letting the host select the maximum, we can use a scheme that will require only $\log_2 P$ steps to find the maximum component. Assume that there are P processors on a d-dimensional hypercube, and each processor P_i has a residual component e_i. To find the maximum of all e_i's, we can use the following scheme on each processor P_i:

Verification Algorithm 2

do $j = 1,\ d$

    1. Processor P_i sends e_i to processor P_j and receives e_j from P_j. The binary IDs of P_i and P_j differ only in the jth digit.

    2. $e_i = max\{e_i,\ e_j\}$

end do

The variable e_i is the maximum value of $\{e_1,\ e_2 \cdots e_p\}$ after the loop is completed. Two important observations can be made about this scheme.

- Only $d = \log_2 P$ steps are needed to complete the process. All communications are between the nearest neighbors. The P processors can be paired into $P/2$ pairs and exchange e_i's simultaneously at each step.

- The final maximum value is available on all P processors after the loop is completed. Therefore all processors can determine independently whether the convergence criterion has been satisfied, thus saving another round of communications used by the host to notify all processors to either stop or continue the iterations.

Fig. 4.14 illustrates the comparison and selection process on a three-dimensional

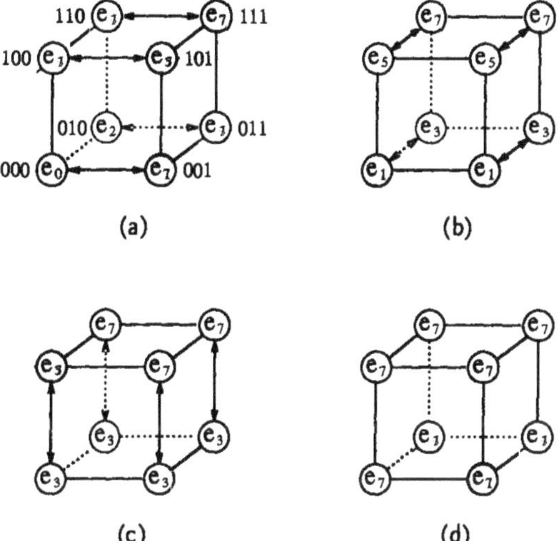

Figure 4.14: (a) Initial distribution of residual components. Processors connected by arrows exchange and compare their residual components. (b) After one comparison step. (c) After two comparison steps. (d) After three comparison steps.

hypercube with $2^3 = 8$ processors. Each processor P_i $(i = 0, 1, \cdots, 7)$ initially has a residual component e_i with

$$e_0 < e_1 < e_2 \cdots < e_7.$$

Three steps are needed to select the largest component. In each step, processors are paired by the arrows as shown in Fig. 4.14 and exchange their data. On each processor, the larger one of the two components is kept after each comparison. As can be seen in Fig. 4.14, the largest component e_7 is available on all processors after three communication steps. This scheme is also implemented as library subroutines on iPSC/860.

Table 4.3 shows the timing results obtained by using the improved commu-

Table 4.3: Timing results of simulation Case 2 on iPSC/860
with improved scheme for verifying
the convergence criterion.
P is the number of processors,
n is the grid size.

n \ P	1	2	4	8	16	32
128×128	21.7	11.1	5.7	3.1	2.0	1.5
128×384	65.2	32.7	16.5	8.5	4.5	3.3

nication scheme for verifying the convergence criterion. All parameters and algorithms are the same as that used in Case 1, except for the improved scheme for verifying the convergence criterion. Comparing the numbers in Table 4.3 with those in Table 4.2, we can see that the computation time is reduced. The improvement is not significant for a small number of processors, like 2 or 4 processors, since in these cases the numerical computation is much more dominant than the communication. As more processors are used, the inter-processor communication becomes more dominant. Thus, the improvement of using better scheme for verifying the convergence criterion is more significant with large number of processors and small problem sizes, like solving the 128×128 model on 32 processors. Fig. 4.15 shows the speedup curves for the data given in Table 4.3. It is obvious that the speedup curves in Fig. 4.15 show better performance than those in Fig. 4.13.

Case 3. As discussed before on page 153, there are three places where inter-processor communication is needed.

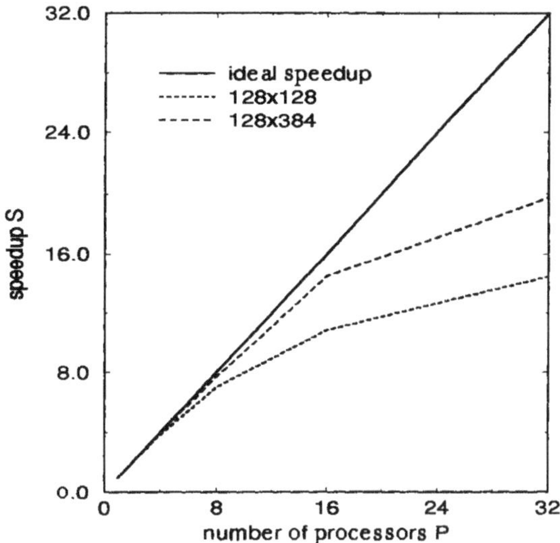

Figure 4.15: Speedup curves for simulation Case 2 on iPSC/860.

- Verification of convergence criterion
- Passing solutions of red lines on the boundary of subdomains
- Passing solutions of black lines on the boundary of subdomains

In simulation Case 2, we used a more efficient communication scheme for verifying the convergence criterion. The new scheme takes advantage of high connectivity on a hypercube and reduces the communication steps from $O(P)$ to $O(\log_2 P)$. We now discuss improvement for the other two communication steps by overlapping communications with computations.

To simplify the discussion, let us assume that there are totally P processors available for solving Eqs. (4.10) and (4.11), the domain is a rectangular two-dimensional grid with even numbers of grid lines in both x and y directions. Each processor is assigned a subdomain, and each subdomain starts with a red line and ends with a black line similar to the subdomains shown in Fig. 4.10 and 4.11. The improvement for communication is based on the fact that all equations in Eqs. (4.10) and (4.11) are independent to each other and can be solved in any order respectively. The new flow chart of the program execution is given in Fig. 4.16. Comparing Fig. 4.16 with Fig. 4.12, we can see that the major changes are the use of asynchronous message passing subroutines and the message waiting check points.

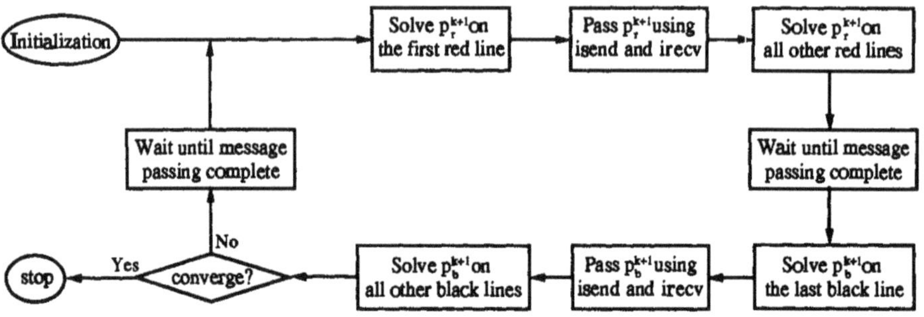

Figure 4.16: Flow chart of program execution with overlapping of communications with computations.

Since all processors, except the first one in the chain, have to pass the solution p_r^{k+1} of the first red line to their predecessor as shown in Fig. 4.10 for computing solutions on the black lines, each processor can calculate the solution p_r^{k+1} of the first red line and then pass it to the predecessor before calculating solutions on the other red lines. Similarly, since all processors, except the last one, have to receive p_r^{k+1} from their successors for calculating solutions on the black lines, each processor (except the last one) should get ready for the incoming message before calculating solutions on the other red lines. The sending and receiving of p_r^{k+1} can be accomplished by using the subroutines isend and irecv. Since these two subroutines are non-blocking, the program can continue the computations of solutions on the other $l-1$ red lines, regardless whether p_r^{k+1} has been sent to the predecessor and received from the successor. In this way, computations are overlapped with communications. After all solutions on the l red lines have been calculated, we need to check to ensure that the message passing has completed. If the message has not arrived, the computations for the solutions on the black lines can not start.

Similar to the situation on red lines, the communications of p_b^{k+1} can also be overlapped with the computations on black lines. Since all processors, except the last one, have to send p_b^{k+1} on the last black line to their successors as shown in Fig. 4.11, a processor should first compute solution p_b^{k+1} on the last black line, then send it to the successor and get ready to receive p_b^{k+1} from the predecessor using non-blocking isend and irecv respectively. The communication processes

are overlapped with the computations of solutions on the other $l-1$ black lines.
Table 4.4 shows the timing results for simulation Case 3 which overlaps commu-

Table 4.4: Timing results of simulation Case 3 on iPSC/860
with improved communication scheme for verifying
the convergence criterion and the overlapping of
communications with computations.
P is the number of processors,
n is the grid size.

n＼P	1	2	4	8	16	32
128×128	22.1	10.5	5.3	2.7	1.5	1.2
128×384	65.2	31.3	15.7	8.0	4.2	2.5

nications with computations. The corresponding speedup curves are given in
Fig. 4.17. It is obvious that the execution time is further reduced as compared

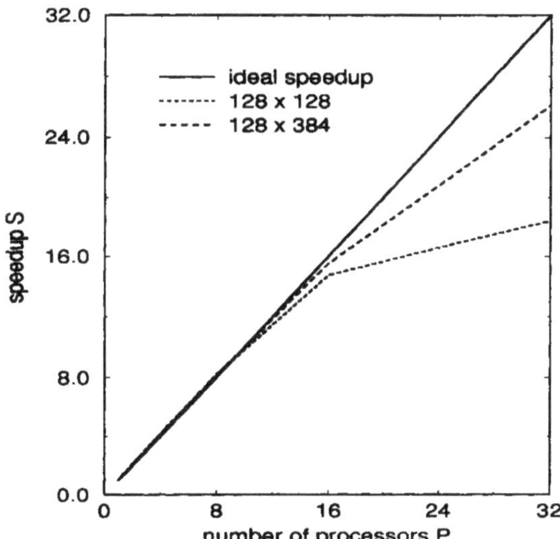

Figure 4.17: Speedup curves for simulation Case 3 on iPSC/860.

to the results given in Tables 4.2 and 4.3. Furthermore, the improvement is
clear for the number of processors ranging from 2 to 32. It is interesting to see
that when the number of processors increases from 1 to 2, the execution time is

reduced by a factor of more than 2. Similar results which are sometimes called "superlinear speedup" have also been reported in other cases [72, 81, 146, 201]. The explanation varys depending on the problem being solved and the architecture used in the computation. For iPSC/860 hypercubes, as discussed in Section 3.1, each processor has 8 Kbytes of high speed cache. With two processors, the size of cache is doubled, and the communication between the two processors can be completely overlapped with the computation if the problem being solved is large enough. Therefore, the aggregate computational speedup is increased by a factor of more than two with two processors due to the double-sized cache. As more processors are used, although the total size of cache increases proportionally, the communication complexity also increases. If two processors are used in the computation, each processor needs to communicate with only one processor. However, if more than two processors are used, a processor not in the either end of the chain has to send a message to one processor and receive a message from another processor, as shown in Fig. 4.10 and Fig. 4.11. Additionally, the amount of computation on each node decreases as more processors are used in the computation, making it harder or impossible to completely overlap communications with computations. The net effect is that the gain from the increased amount of cache is not enough to offset the lose due to the complicated interprocessor communications. Therefore, the speedup curves eventually deviates from the ideal speedup curve as more processors are used in the computation.

Overall, if we examine Tables 4.2, 4.3 and 4.4, and Figs. 4.9, 4.11, and 4.13, we can see clearly the improvement in algorithm performance. With the same algorithm and same computer, a betten implementation which takes advantage of a specific architecture can produce much better speedup.

4.4.2 Results on KSR-1 Parallel Computers

The architecture and programming environment of KSR-1 have been discussed in Sections 3.5 and 3.6. Similar to the discussion for iPSC/860, we also discuss three simulation cases here on KSR-1 for solving the same reservoir simulation problem as given in the previous section. Case 1 is the results obtained by using the completely automatic parallelization strategy. Case 2 shows the results obtained by using semi-automatic parallelization which specifies where to parallelize. Finally, Case 3 presents the results obtained by using semi-automatic parallelization to specify both where to parallelize and the granularity of the parallel tasks. For more detailed discussion on automatic and semi-automatic parallelizations, readers are referred to Sections 3.5 and 3.6.

Case 1. In this case, automatic parallelization is used to parallelize the reservoir simulation algorithm. The algorithm was implemented in Fortran 77, and the parallelization preprocessor KAP was used to parallelize the loops in the code. The compilation process is shown in Fig. 3.23. Only one team of pthreads was used. The number of pthreads in the team is the same as the number of processors.

Table 4.5 shows the timing results for different grid sizes and numbers of pro-

Table 4.5: Timing results of simulation Case 1 on KSR-1
using automatic parallelization.
P is the number of processors,
n is the grid size.

n＼p	1	3	4	5	7	16	32
64 × 64	6.56	10.1	10.2	10.2	10.5	12.6	14.7
128 × 128	244	118	89.9	89.6	77.7	70.7	77.2
256 × 256	2093	817	627	566	446	281	303

cessors used. The times reported here are for advancing the solution over one time step. The output report of automatic parallelization process shows that the KAP parallelized the code at a very fine grain level (matrix-vector product, vector update, etc.), rather than the coarse grain level as indicated by Eqs. (4.10) and (4.11), and the reservoir simulation algorithm given on page 153. From Eqs. (4.10) and (4.11), it is clear that all equations in each of the group can be solved in parallel. Since there are many subroutine calls involved in forming and solving Eqs. (4.10) and (4.11), the KAP is not able to figure out the data dependence and parallelize the code at the coarse grain level. The numbers in Table 4.5 indicate that the 64×64 model appears to be too small to gain anything from parallel processing. The 128×128 and 256×256 models are better, but still far from satisfactory. We see that the best speedup is about 3.5 for the 128×128 model and 7.5 for the 256×256 model, both were achieved on 16 processors. In [202], much better speedup has been reported for solving a regularized least squares problem using the automatic parallelization preprocessor KAP on KSR-1. If we analyze these two problems in terms of the degree of parallelism, the reservoir simulation problem actually appears to be more parallelizable than the regularized least squares problem. In the reservoir simulation algorithm given on page 153, Eqs. (4.10) and (4.11) can all be formed and solved in parallel in each iteration, which makes the load balance and data distribution easy. While for the regularized least squares problem, the Householder Transformation uses the standard ijk loops to factorize the coefficient matrix. The i and j loops carry less and less work as the computation progresses, which makes the load among different processors unbalanced and causes data exchange among different processors. The explanation for the poor performance of the automatically parallelized reservoir simulation algorithm is that since the reservoir simulation code is much more complicated (several thousand lines) than the code for solving the regularized least squares problem (a few hundred lines), the automatic parallelization tool KAP is not able to detect the parallelism at coarse grain level. As a result, KAP parallelizes all loops at the very fine grain level where no data dependence is detected. Many of these loops do not carry enough work, so the gain from parallel processing is not enough to cover the parallel overhead and the additional data movement for distributing data across different processors.

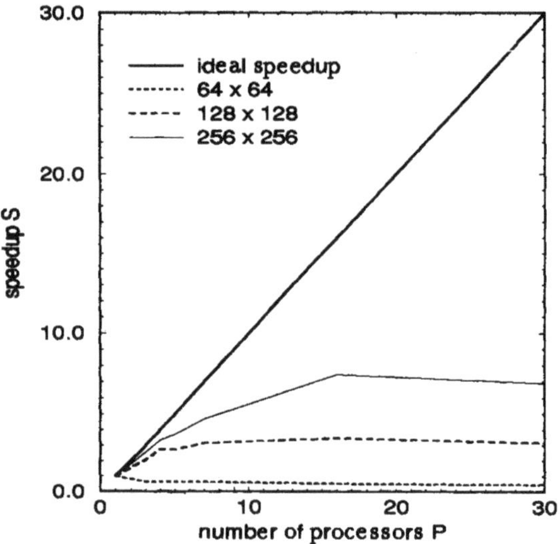

Figure 4.18: Speedup curves for the reservoir simulation Case 1 on KSR-1 parallel computers.

Case 2. With KSR Fortran, a programmer can decide where and how to parallelize a sequential code. As discussed in Case 1, completely automatic parallelization does not always produce satisfactory performance, particularly for large and complicated codes. The major problems involved are

- Where to parallelize. For a large code with thousands of lines, not all loops at the low level should be parallelized, especially if the loop does not encompass significant amount of work.

- How to parallelize. If a section of a code is parallelized, there are many parameters to be determined, like the number of pthreads used to execute the parallel tasks and the granularity of the parallelized tasks.

With automatic parallelization, all these are determined by the KSR compiler and the KAP preprocessor. If a programmer wants more control over where and how the code is parallelized to obtain better performance, he can use the KSR compiler directives and the library subroutines to parallelize a code in a specific way.

The timing results obtained by specifying where to parallelize using semi-automat parallelization are given here. The results obtained by specifying additionally the tile size (see Section 3.6.2.2) will be given in simulation Case 3.

As shown in the reservoir simulation algorithm on page 153, the major part of the computation in reservoir simulation is to form and solve all equations in Eqs. (4.10) and (4.11). The loops represented by the numbers 1, 2, and 3 in the algorithm on page 153 can all be parallelized directly, which produces parallel tasks at very coarse grain level. The "tile" parallel construct (see Section 3.6.2.1) is used here to parallelize the three i-loops in the algorithm. Three KSR compiler directives were used for this purpose.

Table 4.6 shows the timing results for simulation Case 2 on KSR-1 parallel computers. Since it has been shown in Case 1 that the 64×64 model is too small to gain anything from parallel processing, we removed it from further testing and added a larger 256×512 model in the numerical experiments. If we compare the timing results for the 128×128 and 256×256 models in Table 4.5 and Table 4.6, we can see that the execution time is significantly reduced for the same problem size and same number of processors. Even in the single processor case, we see that the results in Table 4.6 is better than that in Table 4.5. This is due to the fact that in Case 1, automatic parallelization by KAP parallelized too many loops at very fine grain level and introduced too much parallel processing overhead, which makes the code less efficient even if the code is running on just one processor.

Table 4.6: Timing results of simulation Case 2 on KSR-1
specifying where to parallelize.
P is the number of processors,
n is the grid size.

n \ P	1	3	4	5	7	16	30
128×128	232	81.6	58.2	58.4	58.3	58	58
256×256	2068	709	516	493	512	256	257
256×512	4213	1416	1057	660	273	259	857

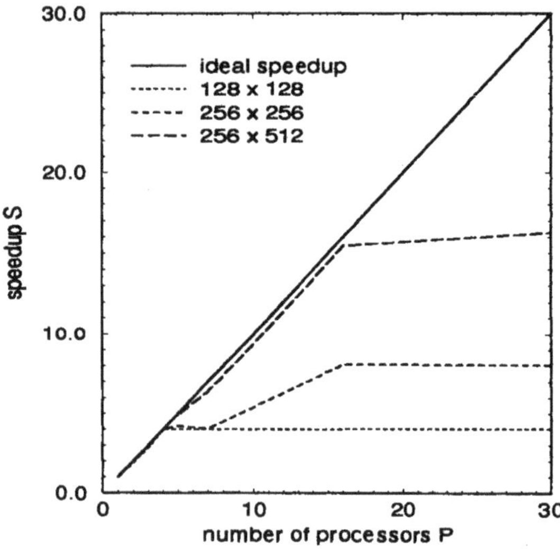

Figure 4.19: Speedup curves for the reservoir simulation Case 2 on KSR-1
parallel computers.

Fig. 4.19 shows the speedup curves for the data in Table 4.6. Although the execution time of simulation Case 2 is better than that of Case 1, Fig. 4.19 does not show much improvement in speedup and scalability for simulation Case 2 when more processors are used in the computation.

After extensive numerical experiment, it was found that the default *tilesize* (see Section 3.6.2.2) specified by the KSR compiler is not properly adjusted based on the bounds of loop indices, the number of processors, and the amount of work in each iteration of the loop.

Case 3. Based on the results and analysis of simulation Case 2, we decided to specify the tile size explicitly in the program, rather than using the default `tilesize` determined by the compiler. To reduce the parallel processing overhead and the data movement between different processors, the tile size of loop 1 in the reservoir simulation algorithm of page 153 is set as

$$tilesize = \left\lceil \frac{2m + P - 1}{P} \right\rceil, \tag{4.12}$$

and that of loop 2 and loop 3 in the same algorithm is set as

$$tilesize = \left\lceil \frac{m + P - 1}{P} \right\rceil, \tag{4.13}$$

where $2m$ is the number of vertical (or horizontal) grid lines, m is the number of red or black lines, P is the number of processors used, and the brackets in Eqs. (4.12) and (4.13) represent chopping the result to obtain an integer if the fraction is not an integer.

The tile sizes given by (4.12) and (4.13) generate, for each loop, the same number of tiles as the number of available processors with the best possible load distribution. Although the "slice" strategy is supposed to produce tiles for a given loop with the same tile sizes as given in (4.12) and (4.13) [108], numerical experiments have shown that the execution is much slower when the slice strategy, rather than the `tilesize`, is specified in the tile parallel constructs.

Table 4.7 contains the timing results obtained by specifying that the three loops in the reservoir simulation algorithm given on page 153 should be parallelized with the tile sizes given by (4.12) and (4.13). The code for simulation Case 3 differs from that for Case 2 only in that the `tilesize` parameter is specified in the three tile constructs.

It is clear from Table 4.7 that, by just specifying three `tilesize` parameters in the three tile constructs of the reservoir simulation algorithm, the execution

Table 4.7: Timing results of simulation Case 3 on KSR-1
specifying where to parallelize and the tilesize.
P is the number of processors,
n is the grid size.

n \ P	1	3	4	5	7	16	30
128×128	232	80.6	58.2	49.0	36.4	14.8	10.1
256×256	2068	696	516	445	304	129	76.0
256×512	4218	1417	1055	857	611	273	149

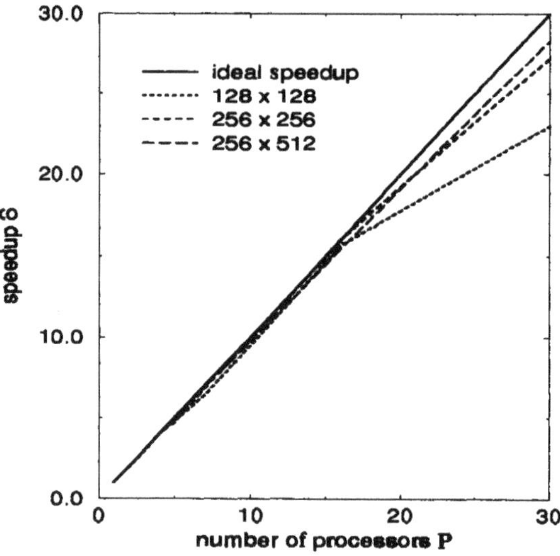

Figure 4.20: Speedup curves for the reservoir simulation Case 3 on KSR-1
parallel computers.

times are significantly reduced compared to those given in Tables 4.5 and 4.6. The code also shows much better scalability in performance as more processors are used in the computation. Fig. 4.20 shows the speedup curves for the data given in Table 4.7. We see that the achieved speedups are very close to the ideal speedup and significantly improved as compared with those given in Figs. 4.18 and 4.19.

The three simulation cases discussed here shows that the code performance can be significantly improved by using parallel constructs and specifying relevant parameters properly. The program implementation is significantly simplified with a simple unified address space on KSR-1.

Chapter

5

Parallel Time Stepping Algorithms

In this chapter, we will discuss a more general approach to the parallelization of numerical algorithms for solving parabolic partial differential equations. The central idea is to exploit parallelism in both the temporal dimension and the spatial dimensions, so that different processors can work on solutions at different time levels simultaneously.

5.1   A New Dimension for Parallelization

Since there is always a spatial domain involved in the numerical solutions of partial differential equations, the major effort in parallelizing the solution processes has been concentrated on spatial domain splitting (domain decomposition), in which the original spatial domain is decomposed into subdomains and different processors work on different subdomains.

Many researchers, however, have observed that the efficiency of the iterative algorithms used for solving those PDEs degrades significantly as the number of processors increases on parallel computers [198], especially on architectures with fast numerical computation processors but relatively slow inter-processor communication channels. For time dependent PDEs, while there are equations to be solved at thousands of time steps, all processor can work only at the same time step at any given moment with the traditional domain splitting, since the computation in the temporal dimension has been treated sequentially. As a result, the communication and computation ratio is very high for large number of processors. A new approach to alleviate this problem is to develop algorithms exploiting parallelism in both the spatial and the temporal dimensions, so that different processors can work on different time steps simultaneously.

To simplify the presentation, the equation

$$\frac{\partial u}{\partial t} + Lu = f \qquad \underline{x} \in \Omega \qquad t_0 < t < T \tag{5.1}$$

$$B(u) = u_b(t) \qquad \underline{x} \in \partial\Omega \qquad t_0 < t < T$$

$$u(\underline{x}, t_0) = u_0(\underline{x}) \qquad \underline{x} \in \Omega$$

is used in most of the sections of this chapter to demonstrate the solution process and the existing problems affecting the performance of scalable parallel computers. L is an elliptic spatial operator (for example L will be a Laplacian if equation (5.1) is the heat equation), B is the boundary operator, and Ω is a rectangular two-dimensional spatial domain. The method presented here can be expanded to more complicated situations involving higher order temporal derivatives and three-dimensional spatial domains.

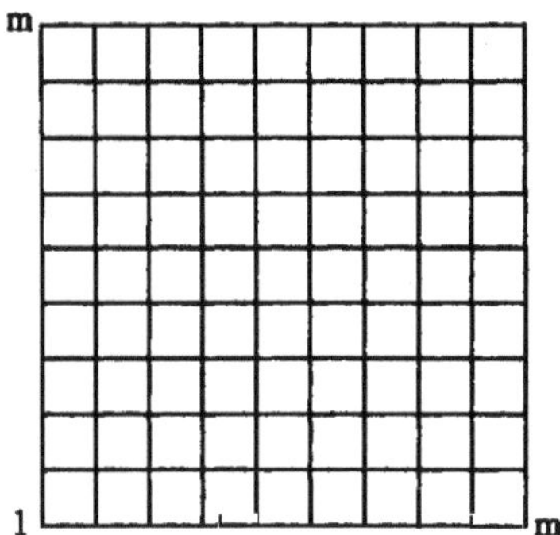

Figure 5.1: a two-dimensional spatial domain

As discussed before, the first step for the numerical solutions of PDEs is to discretize the equations and the domains (in both the spatial and the temporal dimensions). The discretized spatial domain is represented by a set of points in an $m \times m$ grid as shown in Fig. 5.1, and the temporal domain is represented by discrete time steps $t_0 = t_1 < t_2 < \cdots < t_n = T$. If a two-level implicit time stepping algorithm is used, we will have, at each time step t_k, an algebraic equation system

$$Au^k = Cu^{k-1} + b_k \qquad k = 1, 2, \cdots, n \tag{5.2}$$

where A and C are $m^2 \times m^2$ matrices whose entries depend on the forms of the operators L and B and the discretizing scheme, m^2 is the number of grid points in the discrete spatial domain, u^k is an $m^2 \times 1$ vector containing function values u of all grid points at time step k, and b is the discrete form of f. The central part of the solution process is to solve equation (5.2) defined on the discrete spatial domain at each time step t_k.

From equation (5.2), we can see that the equation systems with different index k seem to be dependent on each other and can not be solved simultaneously. Starting from u^0 which is known from the initial condition, we can solve u^1 first, then u^2 can be solved by putting u^1 in the right hand side of equation (5.2). The process repeats until $u^1, \cdots, u^n$ have all been solved. This kind of time stepping process seems to be inherently sequential. Therefore, the major effort in parallelizing the solution process has been concentrated on exploiting parallelism by distributing columns or rows of matrix A to different processors. Since each row in A represents an equation at a spatial grid point, the approach is equivalent to splitting the original spatial domain into subdomains and assigning different subdomains to P different processors as shown in Fig. 5.2. In Fig. 5.2(a), the original $m \times m$ domain is divided into P

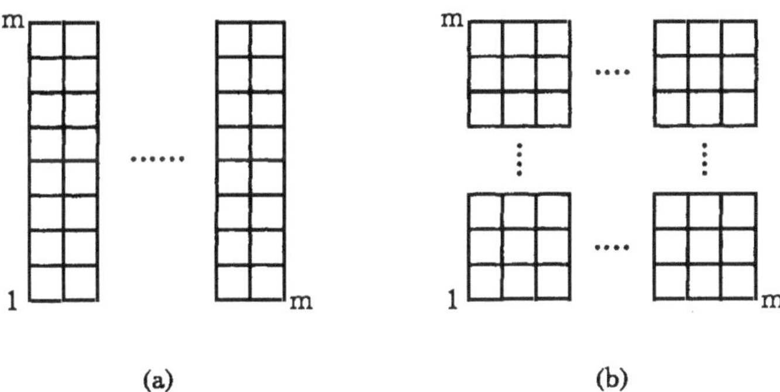

(a)   (b)

Figure 5.2: two domain decomposition schemes

strips with dimensions $m \times \frac{m}{P}$, while in Fig. 5.2(b) the original domain is divided into P patches with dimensions $\frac{m}{\sqrt{P}} \times \frac{m}{\sqrt{P}}$.

Since matrix A is usually sparse and structured, iterative algorithms are used most of the time to solve equation (5.2), such as the Jacobi or Gauss-Seidel type relaxation schemes:

$$u_i^k = T u_{i-1}^k + P(b_k + u^{k-1}) \qquad i = 1, 2, \cdots, \qquad k = 1, \cdots, n.$$

Many iterations are needed at each time step. In each iteration, a processor must exchange solution values on the boundary of its subdomain with processors holding adjacent subdomains by interprocessor communications.

For a given spatial domain, like the one in Fig. 5.1, the larger the number of available processors on a parallel computer, the smaller the size of the subdomain assigned to each processor. Since a processor must exchange solutions on the boundary of its subdomain with processors holding adjacent subdomains, the communication/computation ratio will increase as the number of processors increases. An increasing portion of time will be spent on exchanging information, rather than on computing. Since the time spent on the interprocessor communication does not contribute directly to the numerical computation, the efficiency of the algorithm will degrade as the number of processors increases. Although the overall amount of numerical computation is tremendous because Eq. (5.2) has to be solved at many time steps, all processors can work only at the same time step at any given moment with the sequential time stepping algorithm. Each processor works on a small portion of A and communicates after each iteration with other processors holding adjacent subdomains.

If we look at all equations at different time steps in Eq. (5.2) from a global point of view, it is obvious from Fig. 5.3 that different problems at different time steps can be considered as a combined problem defined on an augmented spatial-temporal domain. Fig. 5.3(a) shows a sequence of n problems defined at n time steps on a two-dimensional spatial domain D. These equations are solved one after another sequentially in the traditional time stepping algorithms. Fig. 5.3(b) shows that the n problems are merged together as a larger system of equations defined on a three-dimensional augmented spatial-temporal domain Ω. Different parallel algorithms for solving systems of algebraic equations can then be applied to the augmented system, so that the solutions at different time steps can be solved in parallel.

Several different approaches to exploiting parallelism in the time dimension will be given in the next few sections, including the waveform relaxation which is based on the ordinary differential equation formulation, and the window and pipeline iterations which are based on the partial differential equation formulation.

5.2 Waveform Relaxation

Waveform Relaxation algorithm was originally developed for solving large systems of ordinary differential equations arising from electric circuit simulations [119, 120, 139, 187, 188]. Since a partial differential equation can be first discretized in the

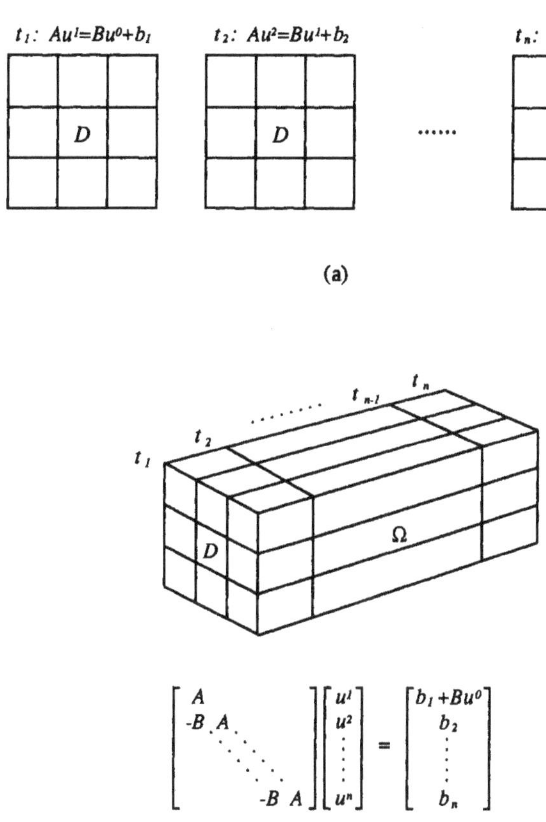

Figure 5.3: (a) A sequence of two-dimensional problems at n time steps defined on domain D. (b) The n two-dimensional problems are combined into a larger problem defined on an augmented spatial-temporal domain Ω.

spatial variables with the time variable remaining continuous, thus generating a set of ordinary differential equations, we can also apply the algorithms developed for ODEs to time dependent PDEs [176]–[181].

5.2.1 Waveform Relaxation for Time Dependent PDEs

Consider the parabolic equation given by Eq. (5.1), we can first discretize the spatial operator L to obtain a system of ordinary differential equations

$$\frac{d\mathbf{u}}{dt} + A(\mathbf{u}) = \mathbf{f} \quad t_0 < t < T \tag{5.3}$$

$$\mathbf{u}(t_o) = \mathbf{u}_0$$

where $\mathbf{u}$ is a vector containing values of the unknown function $u(t)$ at all grid points, A is the discretized version of operator L (in the linear case, A will be a constant matrix), $\mathbf{f}$ is the forcing term determined by function f and the boundary conditions in Eq. (5.1), and $\mathbf{u}_0$ is the initial condition.

As a simple example, if the spatial derivatives of the following equation

$$\frac{\partial u}{\partial t} - \frac{\partial^2 u}{\partial x^2} = f(x,t) \qquad x \in \Omega \quad t_o < t < T \tag{5.4}$$

$$u(x,t) = 0 \qquad x \in \partial\Omega \quad t_o < t < T$$

$$u(x,t_0) = g(x) \qquad x \in \Omega$$

are discretized on a uniform grid using the central finite difference scheme, we will have a system of ordinary differential equations similar to Eq. (5.3) with

$$\mathbf{u}(t) = \{u_1(t), \cdots, u_N(t)\}^T,$$

$$\mathbf{f}(t) = \{f_1(t), \cdots, f_N(t)\}^T,$$

$$A = \begin{bmatrix} -2 & -1 & & \\ -1 & \ddots & \ddots & \\ & \ddots & \ddots & -1 \\ & & -1 & -2 \end{bmatrix},$$

where N is the number of grid points. The waveform relaxation algorithm for solving Eq. (5.3) is very similar to that of the classic relaxation algorithms discussed in Section 2.2 for solving algebraic equations. With the waveform Jacobi relaxation, Eq. (5.3) becomes

$$\frac{du_i^{n+1}}{dt} + A(u_1^n, u_2^n, \cdots, u_i^{n+1}, \cdots u_N^n) = f_i, \tag{5.5}$$

$$i = 1, 2, \cdots N, \quad n = 0, 1, 2, \cdots.$$

Starting from a given initial value vector, the N ordinary differential equations are independent to each other and can be solved using any standard algorithms and packages for ordinary differential equations [61, 62, 116]. For the equation given by Eq. (5.4) and a uniform grid, the waveform Jacobi relaxation will simply be

$$\frac{du_i^{n+1}}{dt} - \frac{1}{\Delta x^2}(u_{i-1}^n - 2u_i^{n+1} + u_{i+1}^n) = f_i, \tag{5.6}$$

$$i = 1, 2, \cdots N, \quad n = 0, 1, \cdots.$$

where Δx is the grid spacing. If the implicit Euler's algorithm is used to intergrate Eq. (5.6), we will have

$$\frac{u_{i,k}^{n+1} - u_{i,k-1}^{n+1}}{\Delta t} - \frac{1}{\Delta x^2}(u_{i-1,k}^n - 2u_{i,k}^{n+1} + u_{i+1,k}^n) = f_{i,k},$$

$$i = 1, 2, \cdots N, \quad k = 1, 2, \cdots m, \quad n = 1, 2, \cdots,$$

or equivalently as

$$-u_{i,k-1}^{n+1} + (1 + 2\lambda)u_{i,k}^{n+1} = \lambda u_{i-1,k}^n + \lambda u_{i+1,k}^n + \Delta t f_{i,k} \tag{5.7}$$

$$i = 1, 2, \cdots N, \quad k = 1, 2, \cdots m, \quad n = 0, 1, \cdots,$$

where m is the number of time steps and $\lambda = \Delta t/\Delta x^2$. Upon assembling all equations in Eq. (5.7) with respect to index k (the time step), we obtain a matrix equation system at each grid point x_i

$$Bu_i^{n+1} = \lambda u_{i-1}^n + \lambda u_{i+1}^n + \Delta t f_i, \tag{5.8}$$

$$i = 1, 2, \cdots N, \quad n = 0, 1, \cdots,$$

where

$$B = \begin{bmatrix} (1+2\lambda) & & & \\ -1 & (1+2\lambda) & & \\ & \ddots & \ddots & \\ & & -1 & (1+2\lambda) \end{bmatrix},$$

$$u_i^n = (u_{i,1}^n, u_{i,2}^n, \cdots, u_{i,m}^n)^T,$$

$$f_i = (f_{i,1} + \frac{g_i}{\Delta t}, f_{i,2}, \cdots, f_{i,m})^T.$$

Note that g_i is calculated from the initial condition given in Eq. (5.4). The complete algorithm for solving Eq. (5.5) can then be given as:

Algorithm 5.1
Waveform Jacobi Relaxation

$n = 0$
select initial guess $u_i^0(t)$
$i = 1, 2, \cdots N, \quad t_0 < t < T$

do until converge
    for $i = 1, N$
        solve Eq. (5.5) with $u_i^{(n+1)}(t_0) = u_{0i}$
    end for
    $n = n + 1$
end do

Note that in the above algorithm, the initial guess $u_i^0(t)$ is a function of time t at grid point i. If the integration of ODEs has been prescribed at a set of discrete time steps $[t_1, t_2, \cdots t_k]$, then $u_i^0(t)$ can also be a vector of the form $[u_i^0(t_1), u_i^0(t_2), \cdots u_i^0(t_n)]^T$. The value $u_i^0(t_0)$, as well as $u_i^{n+1}(t_0)$, should be the initial value given in Eq. (5.4) at a proper grid point, i.e.

$$u_i^0(t_0) = u_i^{n+1}(t_0) = g(x_i) = u_{0i},$$

$$i = 1, 2, \cdots N, \quad n = 0, 1, \cdots.$$

Similar to the classic relaxation algorithms, we can use the waveform Gauss-Seidel and waveform SOR relaxations to solve Eq. (5.3) more efficiently. Based on Eq. (5.5), the waveform Gauss-Seidel relaxation can be written as

$$\frac{du_i^{n+1}}{dt} + A(u_1^{n+1}, \cdots, u_{i-1}^{n+1}, u_i^{n+1}, u_{i+1}^n, \cdots, u_N^n) = F_i, \tag{5.9}$$

$$i = 1, 2, \cdots N, \quad n = 0, 1, \cdots.$$

where the most recently updated functions $u_1^{n+1}(t), u_2^{n+1}(t), \cdots, u_{i-1}^{n+1}(t)$, are used in the solution process of $u_i^{n+1}(t)$. With this relaxation algorithm, the heat equation defined on a two-dimensional rectangular domain can be discretized as

$$\frac{du_{i,j}^{n+1}}{dt} - \frac{1}{h^2}(u_{i,j-1}^{n+1} + u_{i-1,j}^{n+1} - 4u_{i,j}^{n+1} + u_{i+1,j}^n + u_{i,j+1}^n) = f_{i,j}. \tag{5.10}$$

$$i,j = 1, 2, \cdots m, \quad n = 0, 1, \cdots.$$

Similarly, the waveform SOR relaxation can be written as

$$\frac{d\hat{u}_i}{dt} + A(u_1^{n+1}, \cdots, u_{i-1}^{n+1}, \hat{u}_i, u_{i+1}^n, \cdots, u_N^n) = F_i, \tag{5.11}$$

$$u_i^{n+1} = (1 - \omega)u_i^n + \omega\hat{u}_i$$

$$i = 1, 2, \cdots N, \quad n = 0, 1, \cdots.$$

Unfortunately, the straightforward application of the waveform Jacobi and Gauss-Seidel relaxations to solving parabolic PDEs has not been very successful due to the slow convergence rate [176]. It was shown in [129] that the waveform relaxations based on the Jacobi and Gauss-Seidel iterations converge at a rate of $1 - O(h^2)$ where h is the grid spacing. More detailed convergence analysis for linear and nonlinear systems are given in [130, 135, 136, 137, 138]. Even with the waveform SOR relaxation, the parameter ω does not accelerate the convergence rate significantly as it does for linear algebraic equations with symmetric positive definite coefficient matrices.

It has been discussed in Section 2.4 that the convergence rate of the classic relaxation algorithms can be significantly accelerated using the multigrid method. This naturally leads to the multigrid waveform relaxation.

5.2.2 Multigrid Waveform Relaxation

Comparing Eq. (5.3) with Eq. (2.45) in the discussion of the multigrid algorithm in Section 2.4.1, we see that the difference between the two equations is that Eq. (2.45) represents all linear algebraic equations arising from the discretization in both space and time, with one algebraic equation per grid point, while Eq. (5.3) represents all ordinary differential equations arising from the discretization in space only, with one ordinary differential equation per grid point. The component x_i in the unknown vector of Eq. (2.45) is a scalar number defined at a grid point at a particular time step, while the component $u_i(t)$ in the unknown vector of Eq. (5.3) is a function of time t. If solutions are sought at discrete time steps $(t_1, t_2, \cdots, t_k)$, then the component $u_i(t)$ itself is a vector $(u_i(t_1), u_i(t_2), \cdots, u_i(t_k))$. Therefore, all the basic operations, including smoothing, interpolation, and prolongation, of the multigrid method originally applied to a scalar x_i should now be applied to a function $u_i(t)$ or a vector $(u_i(t_1), u_i(t_2), \cdots, u_i(t_k))$.

The application of the multigrid method to the waveform relaxation for accelerating the convergence rate was first proposed by Lubich and Ostermann [122], and also independently by Vandewalle and Piessens [179]. The procedure is very similar to that for solving linear algebraic equations, except that all multigrid operations are applied to vectors, rather than scalars. The three major operations in the multigrid method for algebraic equations are modified for solving the ODE systems as follows:

- Smoothing. The smoothing operation is essentially one or more Jacobi or Gauss-Seidel type relaxations. With a initial guess of $u_i^0(t)$, the waveform relaxation for linear PDE systems can be written in a unified form as

$$\frac{d\mathbf{u}^{n+1}}{dt} + N\mathbf{u}^{n+1} = P\mathbf{u}^n + \mathbf{f} \tag{5.12}$$

$$n = 0, 1, 2, \cdots$$

where A is split as $A = N - P$. If we write $A = D - L - U$ with D, L, and U the diagonal, the lower triangular and the upper triangular parts of A respectively, We will have:

$$
\begin{aligned}
N = D, \quad & P = L + U & \text{for Jacobi relaxation,} \\
N = D - L, \quad & P = U & \text{for Gauss-Seidel relaxation,} \\
N = \frac{1}{\omega}D - L, \quad & P = \frac{1-\omega}{\omega}D + U & \text{for SOR relaxation.}
\end{aligned}
$$

- Residual. The residual is the vector by which the approximate solution $\mathbf{u}^{n+1}$ fails to satisfy the original equation, i.e.

$$\mathbf{r}^{n+1} = \frac{d\mathbf{u}^{n+1}}{dt} + A(\mathbf{u}^{n+1}) - \mathbf{f}. \tag{5.13}$$

Since the calculated approximate solution after the smoothing operation satisfies Eq. (5.12), it is easy to see that

$$
\begin{aligned}
\mathbf{r}^{n+1} &= \frac{d\mathbf{u}^{n+1}}{dt} + A(\mathbf{u}^{n+1}) - \mathbf{f} \\
&= \frac{d\mathbf{u}^{n+1}}{dt} + (N - P)\mathbf{u}^{n+1} - \mathbf{f} \\
&= \frac{d\mathbf{u}^{n+1}}{dt} + N\mathbf{u}^{n+1} - P\mathbf{u}^{n+1} - \mathbf{f} - P\mathbf{u}^n + P\mathbf{u}^n \\
&= P(\mathbf{u}^n - \mathbf{u}^{n+1}),
\end{aligned}
$$

thus $\mathbf{r}^{n+1}$ can be calculated easily without computing the term $\frac{d\mathbf{u}^{n+1}}{dt}$.

- Projection and interpolation. These two operations are exactly the same as in the case for linear algebraic equations, except that the operations are applied to vectors instead of scalars.

Once the basic operations of the multigrid method are defined, the multigrid waveform relaxation (V-cycle version) is very similar to the algorithm for solving linear algebraic equations given on page 61. We have assumed here that there is a hierarchy of grids $\left\{\Omega^h, \Omega^{2h}, \Omega^{4h}, \cdots\right\}$, with Ω^h being the finest grid on which the numerical solution is desired. In Algorithm 5.2, step 1 means that the equation on the coarsest grid is solved exactly by carrying out the relaxations until convergence, or using some other direct solution method. The relaxations in step 2 and 6 are given in the general form which includes the Jacobi and Gauss-Seidel relaxations.

Similar to the case when multigrid method is applied to the linear algebraic equations, there are also W-cycle and F-cycle algorithms for the waveform relaxations. More detailed discussions on the W-cycle and F-cycle multigrid waveform methods are given in [22, 75]. The extension to nonlinear parabolic problems using multigrid waveform relaxations was given in [176]–[181]. The convergence rate of the waveform relaxation has been significantly accelerated using the multigrid method. For example, in [177] the convergence factor (the factor by which the norm of the residual is reduced in each iteration) using the waveform Jacobi relaxation and the trapezoidal integration scheme for a system of 841 ODEs is reported as 0.99. The ODEs were derived by discretizing the equation

$$\frac{\partial u}{\partial t} = \frac{\partial^2 u}{\partial x^2} + \frac{\partial^2 u}{\partial y^2}$$

$$(x,y) \in \Omega = [0,1] \times [0,1], \quad t > 0$$

on a 31×31 uniform grid. Several hundreds of waveform relaxations are needed to obtain an accurate solution. With multigrid acceleration, the convergence rate was measured as 0.11 and an accurate solution was obtained in a few iterations.

It is also reported in [177] that the waveform relaxation with multigrid acceleration is competitive on sequential computers to other solution methods based on direct PDE formulations, for example the Crank-Nicolson method, and is often better on parallel computers than those algorithms based on the direct PDE formulations.

5.2.3 Parallel Implementation of the Multigrid Waveform Relaxations

The parallelization of the multigrid waveform relaxation algorithm is relatively straight forward based on the discussions given in the previous two sections. Most of the reported work in parallel multigrid waveform relaxation is still implemented in a fashion that treats the time dependent computations sequentially [176, 177, 179, 180]. These implementations are based on domain decompositions of the spatial domain. A processor is assigned a subdomain for parallel processing and is responsible for carrying

Algorithm 5.2
V-cycle Multigrid Waveform Relaxation $MWV^h\left(\mathbf{u}^h, \mathbf{f}^h\right)$

if (Ω^h is the coarsest grid) then

   1. solve $\dfrac{d\mathbf{u}^h}{dt} + A^h(\mathbf{u}^h) = \mathbf{f}^h$ exactly

else

   2. Relax n_1 times on
$$\frac{d\mathbf{u}^h_{n+1}}{dt} + N(\mathbf{u}^h_{n+1}) = P(\mathbf{u}^h_n) + \mathbf{f}^h, \quad n = 0, 1, \cdots, n_1 - 1$$
with $\mathbf{u}^h$ as the initial guess $\mathbf{u}^h_0$

$\mathbf{u}^h = \mathbf{u}^h_{n_1}$

   3. $\mathbf{f}^{2h} = I^{2h}_h \big(\dfrac{d\mathbf{u}^h}{dt} + A^h \mathbf{u}^h - \mathbf{f}^h\big)$

   4. $\mathbf{u}^{2h} = 0$

$\mathbf{u}^{2h} = MWV^{2h}(\mathbf{u}^{2h}, \mathbf{f}^{2h})$

   5. $\mathbf{u}^h = \mathbf{u}^h + I^h_{2h} \mathbf{u}^{2h}$

   6. Relax n_2 times on
$$\frac{d\mathbf{u}^h_{n+1}}{dt} + N(\mathbf{u}^h_{n+1}) = P(\mathbf{u}^h_n) + \mathbf{f}^h, \quad n = 0, 1, \cdots, n_2 - 1$$
with $\mathbf{u}^h$ as the initial guess $\mathbf{u}^h_0$

$\mathbf{u}^h = \mathbf{u}^h_{n_2}$

end if

out basic multigrid operations, including smoothing, computing the residual, projecting and injecting between grids, for the ODEs defined on the grid points in that subdomain. As mentioned before, the only difference between the multigrid waveform relaxation and the conventional multigrid algorithm for solving linear algebraic equations is that all the basic operations in the waveform relaxation apply to a vector (representing solution values of a grid point at a set of discrete time steps), not a scalar. All the parallelization strategies for implementing the multigrid algorithm on parallel computers discussed in Section 2.4 can be used here for the implementation of the multigrid waveform relaxation algorithm.

To introduce parallelism into the time dimension, we needed to parallelize the solution process for solving the ODEs arising from the spatial domain decomposition [192]. At each grid point of a subdomain, there is an ODE to be integrated at discrete time steps. For example, the model equation given in Eq. (5.4) will generate an ODE like Eq. (5.6) at each grid point if the waveform Jacobi relaxation method is used. Eq. (5.6) can be rewritten as

$$\frac{du_i^{n+1}}{dt} = -\frac{2}{\Delta x^2}u_i^{n+1} + g_i \tag{5.14}$$
$$i = 1, 2, \cdots m, \quad n = 0, 1, \cdots$$

where g_i includes the term of f_i in Eq. (5.6) and the effects of all the relevant solution values from the previous iteration. For more general situations Eq. (5.14) can be written in the form of

$$\frac{du}{dt} = a(t)u + g(t) \tag{5.15}$$

where for simplicity we have dropped the relaxation index $n + 1$ and the spatial grid index i. For each relaxation step in the waveform relaxation, there is an ODE like Eq. (5.15) at each grid point to be integrated. The simplest one-step Euler's algorithm gives

$$u_{k+1} = (1 + \Delta t a_{k+1})u_k + \Delta t g_{k+1} \tag{5.16}$$
$$k = 0, 1, \cdots, n_t$$

where n_t is the number of time steps. Traditionally, the computation expressed by Eq. (5.16) has been done sequentially on a single processor. To parallelize the process, one can use either the parallel algorithms for linear recurrence equations [80, 91, 110, 111, 113] or the cyclic reduction algorithm [90, 100, 101, 112] for solving banded lower triangular systems. For the latter, we first need to put Eq. (5.16) in the form of

$$
\begin{bmatrix}
1 & & & & \\
b_1 & 1 & & & \\
 & b_2 & \ddots & & \\
 & & \ddots & 1 & \\
 & & & b_{n_t-1} & 1
\end{bmatrix}
\begin{bmatrix}
u_1 \\
u_2 \\
\vdots \\
u_{n_t-1} \\
u_{n_t}
\end{bmatrix}
=
\begin{bmatrix}
\Delta t g_1 - b_0 u_0 \\
\Delta t g_2 \\
\vdots \\
\Delta t g_{n_t-1} \\
\Delta t g_{n_t}
\end{bmatrix},
\qquad (5.17)
$$

where $b_i = -(1 + \Delta t a_{i+1})$, then use the parallel algorithms for banded triangular systems. In particular, a cyclic reduction algorithm was discussed in Section 1.3.1.

If more sophisticated multi-step algorithms are used for integrating Eq. (5.15), the corresponding linear algebraic equation system will then be in the form of a banded lower triangular system as

$$
\begin{bmatrix}
d_{11} & & & & \\
a_{21} & d_{22} & & & \\
\vdots & a_{32} & \ddots & & \\
a_{k1} & & \ddots & \ddots & \\
 & \ddots & & \ddots & \ddots \\
 & & a_{n_t,n_t-k+1} & a_{n_t,n_t-1} & d_{n_t,n_t}
\end{bmatrix}
\begin{bmatrix}
u_1 \\
u_2 \\
\vdots \\
\vdots \\
\vdots \\
u_{n_t}
\end{bmatrix}
=
\begin{bmatrix}
g_0 \\
g_1 \\
\vdots \\
\vdots \\
\vdots \\
g_{n_t-1}
\end{bmatrix}.
\qquad (5.18)
$$

At each spatial grid point, there is a system like Eq. (5.18) whose order is the same as the number of time steps at which solutions are sought. Again, the cyclic reduction (and other parallel algorithms for the banded lower triangular matrix) can be used to make multiple processors work on Eq. (5.18), so that different solution values at different time steps are solved in parallel. Detailed complexity analysis is given in [193] to demonstrate the optimality of this space-time parallel algorithm.

Fig. 5.4 shows the flow chart of the parallel multigrid waveform relaxation algorithm applied to time dependent PDEs. The first step is the spatial discretization and domain decomposition, after which the spatial domain has been decomposed into subdomains and there is an ODE defined at each grid point of the subdomains. Different processors can be assigned different subdomains for parallel processing. The second step is to distribute the integration of the ODEs over the time dimension to different processors so that different processors can work on solutions at different time steps concurrently.

5.3   Pipeline Iteration

In contrast to the waveform relaxation which is based on the ODE formulation, this space-time parallel method is based on the direct PDE formulation [189]. As shown

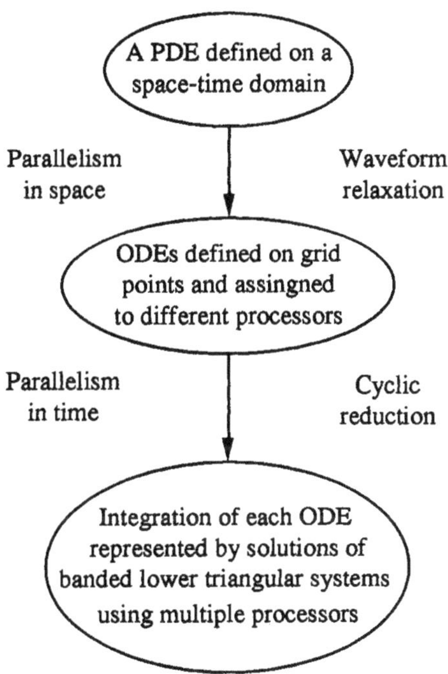

Figure 5.4: Flow chart of the parallel multigrid waveform relaxation algorithm.

by Eq. (5.2), the central part of solving time dependent PDEs numerically is to solve equation

$$Au^k = Cu^{k-1} + b_k, \qquad k = 1, 2, \cdots m, \qquad (5.19)$$

where m is the number of time steps. Since matrix A, which represents the space and time discretization, is usually a sparse matrix, iterative algorithms (like the relaxation methods) are used most of the time to solve Eq. (5.19). The traditional sequential time stepping algorithm can be given as:

Algorithm 5.3
Sequential Time Stepping

Initialize u^0 using the given
initial condition

for $k = 1, m$
    $u_0^k = u^{k-1}$
    $i = 1$
    do until converge
        $u_i^k = Tu_{i-1}^k + P(b_k + u^{k-1})$
        $i = i + 1$
    end do
    $u^k = u_i^k$
end for

Fig. 5.5 shows the flow chart of the sequential time stepping process. Each box represents the iterations at a particular time step. At each step, the converged solution from the previous time step is taken as the initial guess to start the iteration. When the iterations converge, the solution is then passed to the iteration for the next time step as the initial guess. The process continues sequentially until solutions at all desired time steps have been calculated.

We see from Fig. 5.5 that the iterations at time step t_k can not start before the iterations at time step t_{k-1} have converged, since the value u^{k-1} is needed in the right hand side of the equation at time step t_k and is also used as the initial guess u_0^k for the iterations at time step t_k. To simplify the discussion, let us express the iterations at a particular time step t_k as

$$u_i^k = \phi(u_{i-1}^k, u^{k-1}) \qquad (5.20)$$

$$i = 1, 2, \cdots, \quad k = 1, 2, \cdots m.$$

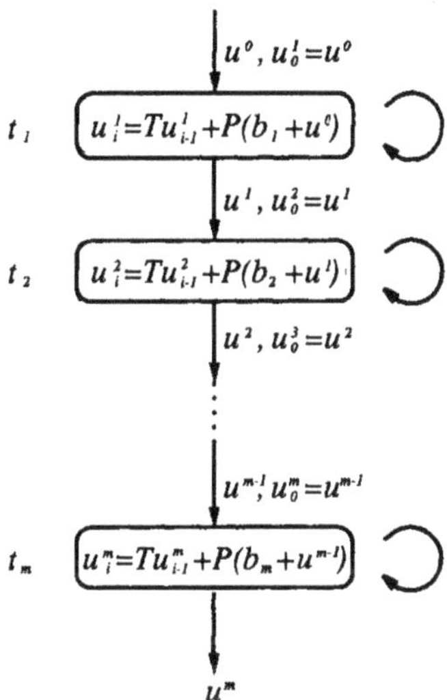

Figure 5.5: Flow chart of the sequential time stepping algorithm. The iteration in each box is executed until convergence before the iteration in the next box can start. Upon convergence at t_k, the solution u^k is used as u_0^{k+1} to start the iteration at t_{k+1}.

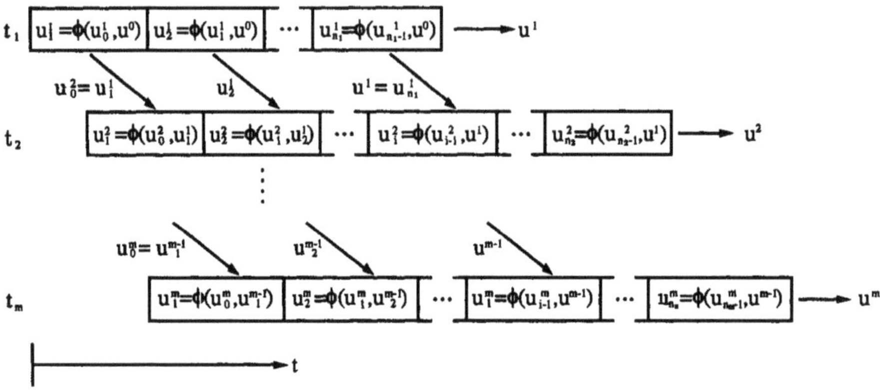

Figure 5.6: Progress of the pipeline iterations in which the intermediate value u_1^{k-1} at time step t_{k-1} is used to start the iterations at time step t_k.

The sequential time stepping algorithm uses the converged value of u^{k-1} as u_0^k to start the iteration given in Eq. (5.20). In the pipeline iteration, however, the iterations at time step t_k starts right after the iteration at time t_{k-1} has generated the value u_1^{k-1}. The value u_1^{k-1} is used as both the initial guess u_0^k to start the iteration at t_k and an approximation for the value u^{k-1} in the right hand side of Eq. (5.20). As the iteration at t_{k-1} goes on, so does the iteration at t_k. In each iteration, the value u_j^{k-1} generated by the iteration at t_{k-1} is passed to the iteration at t_k as a better approximation to u^{k-1} in the right hand side of Eq. (5.20). This process continues until the iteration at t_{k-1} has converged. At this point, the value $u_{n_{k-1}}^{k-1}$ generated by the iteration at t_{k-1} is taken as the right value for u^{k-1} and passed to the iteration at t_k, and the value u_i^k generated by the iteration at time t_k is used to continue the iteration at time t_k. Therefore, after the convergence of the iteration at time t_{k-1}, the parallel time stepping algorithm starts the iteration at time t_k with u_i^k as the initial guess, rather than u^{k-1} which is used as the initial guess in the sequential time stepping. The efficiency of this parallel time stepping algorithm relys on the assumption that u_i^k which is generated concurrently with the computation of u^{k-1} is a better approximation to u^k than is u^{k-1}. Therefore, the remaining iteration at time t_k should converge faster than does the iteration starting with u^{k-1} as the initial approximation to u^k.

Fig. 5.6 shows the progress of this parallel time stepping algorithm. The iterations at different time steps proceed like a pipeline. That is why we call it pipeline iteration in this book. When looking at any two iterations at two consecutive time steps, for example t_1 and t_2, we see that there are three different phases. In the first phase, the iteration at time step t_1 starts and passes u_1^1 to the iteration at t_2. In the second phase, the iterations at both t_1 and t_2 proceed concurrently until the iteration at t_1 has converged. This phase generates, in parallel, the converged solution u^1 at t_1 and a good initial guess u_i^2 for u^2. Finally, in the third phase the iteration at t_2 proceeds until convergence using u_i^2 as an initial guess.

Assume that processor k is assigned Eq. (5.19) defined at a particular time step t_k, the pipeline iteration algorithm on processor k can be given as

Algorithm 5.4
Pipeline Iteration Algorithm

set $i = 0$
if (*not the first processor*) **then**
    receive u_1^{k-1} from processor $k - 1$
    set $u_0^k = u_1^{k-1}$
    do until (*converge on processor $k - 1$*)
        $i = i + 1$
        $u_i^k = \phi(u_{i-1}^k, u_i^{k-1})$
        if (*not the last processor*) send u_i^k to processor $k + 1$
        receive u_i^{k-1} from processor $k - 1$
    end do
end if
set $u^{k-1} = u_i^{k-1}$
set $u_0^k = u_i^k$
$i = 0$
do until (*converge on processor k*)
    $i = i + 1$
    $u_i^k = \phi(u_{i-1}^k, u^{k-1})$
    if (*not the last processor*) send u_i^k to processor $k + 1$
end do
$u^k = u_i^k$

In principle, the two iterations in phase 2 and phase 3 of the pipeline iteration can be of different types, i.e. we can use iteration $u_i^k = \phi_1(u_{i-1}^k, u_i^{k-1})$ to improve the initial guess before the iterations at t_{k-1} converges, and then use iteration $u_i^k = \phi_2(u_{i-1}^k, u^{k-1})$ to get the final solution after the iteration at t_{k-1} has converged. But in practical implementations, ϕ_1 and ϕ_2 are usually the same relaxation scheme [189] for ease of programming.

If there are more time steps than the number of processors, say P processors and n_t time steps with $n_t > P$, the time steps can be assigned to different processors in a wrap around fashion. A particular processor P_i is assigned time steps t_i, t_{i+P}, t_{i+2P}, $\cdots$. The iteration starts at time step t_i, after the convergence, the process continues to time step t_{i+P}. The communication pattern between different processors forms a ring as shown in Fig. 5.7. Note that processor P_P in Fig. 5.7 will not start computation

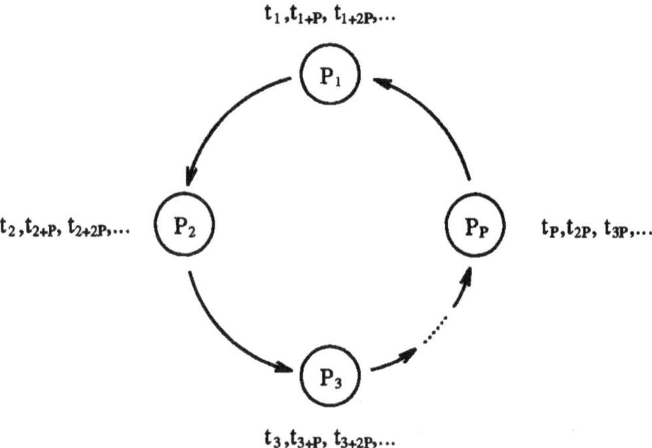

Figure 5.7: Communication pattern (a ring) between different processors in parallel time stepping algorithms

until all previous processors P_1 to P_{P-1} have completed at least one iteration, at which point processor P_1 should have completed $P - 1$ iterations if we assume that it takes the same amount of time to complete one iteration on any of the processors involved. When processor P_1 finishes iterations for computing the solution at time t_1, it will need intermediate solution u_i^P from processor P_P to start the iteration for computing the solution at time t_{1+P}. Depending on whether the iterations at time t_1

converged in less than P iterations, we will have two different situations:

- more than P (P included) iterations, say n iterations with $n > P$, are needed for the iteration at time t_1 to converge. In this case, by the time processor P_1 finishes computing the solution at t_1, processor P_P has finished $n - P + 1$ iterations for computing the solution at time t_P. The larger the value of n, the more the iterations processor P_P has done to produce an initial guess for processor P_1 to start the iteration at time t_{1+P}, this part of the computation at t_p is overlapped with the iteration on P_1 for time step t_1.

- less than P iterations are needed for the iteration at time step t_1 to converge. In this case, by the time processor P_1 finishes computing the solution at t_1, processor P_P has not started the iteration at time t_P yet. Therefore, processor P_1 has to wait $P - n$ iterations before processor P_P can complete one iteration and pass u_1^P to P_1 to start iteration on time step t_{1+P}. The smaller the number of iterations n required for convergence at time step t_1, the longer the waiting period for processor P_1 to start the iteration at the next step t_{1+P}.

Although the above analysis is with respect to two particular processors P_1 and P_P, it is also true for all other processors too. Overall, it is easy to see from the above analysis that this parallel time stepping algorithm has better parallel efficiency when the relaxation scheme used to solve Eq. (5.19) requires more iterations to converge at each time step. This, however, does not mean that efficient or accelerated relaxation algorithms can not be used in combination with this parallel time stepping algorithm. We can use fewer processors for parallel time stepping. As reported in [189], fairly good speed up (as shown in Table 5.1) has been achieved when the parallel time

Table 5.1 : Speedup of the pipeline iteration algorithm with
multigrid relaxation for solving the Burger's Equation
P: number of processors,   S: speedup.

P	1	2	4	8	16
S	1.00	1.96	3.99	7.00	9.38

stepping algorithm with multigrid relaxation is used to solve the Burger's Equation

$$\nu \frac{\partial^2 u}{\partial x^2} - u \frac{\partial u}{\partial x} - \frac{\partial u}{\partial t} = 0, \qquad (x, t) \in (0, 1) \times (0, 5),$$

$$u(x, 0) = \sin(\pi x), \qquad\qquad x \in (0, 1),$$

$$u(0,t) = 0 = u(1,t), \qquad\qquad t \in (0,5),$$

where $\nu = 0.01$. The equation is linearized by lagging the value of u in the coefficient one cycle behind the current value. For details, readers are referred to reference [189].

Just like in the case of the waveform relaxation in which parallelism is exploited at both the spatial and temporal dimensions, the pipeline iteration algorithm discussed in this section can also be implemented in a space-time parallel fashion by decomposing both the space and time domains and assigning the decomposed space-time subdomains to different processors for parallel processing. This is particularly important for good parallel efficiency when a large number of processors are available for the computation. Since it is not always beneficial to use many processors in the time dimension for computing solutions at different time steps concurrently, we need to assign multiple processors to work on the same time step using spatial domain decomposition, so that an iteration completes faster at a particular time step, especially for problems involving large spatial domains (practical engineering models can have millions of grid points in a discretized space domain). Fig. 5.8. shows how 8 pro-

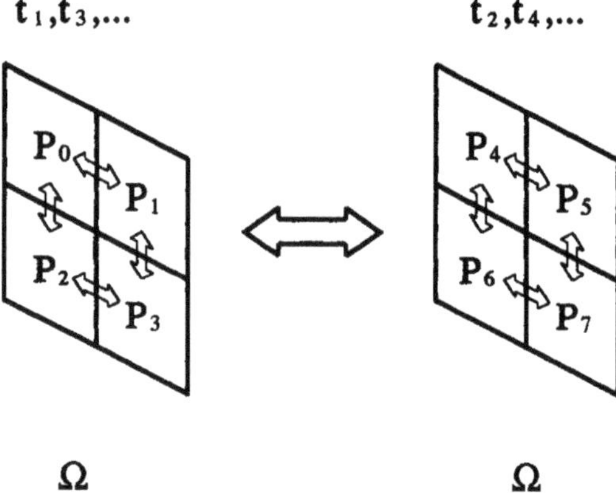

Figure 5.8: Pipeline iteration algorithm implemented with space-time domain decomposition. Arrows indicate inter-processor communications.

cessors are used to solve a PDE defined on a two-dimensional space domain Ω using the pipeline iteration algorithm. We see from Fig. 5.8 that the space domain is decomposed into four subdomains which are then assigned to four different processors.

Processors P_0, P_1, P_2 and P_3 start the first iteration at time t_1 for solving Eq. (5.19). Once u_1^1 is obtained (each of the four processors has only part of u_1^1 corresponding to a particular subdomain), the values are passed to processors P_4, P_5, P_6 and P_7 which then start the iteration at time step t_2. The arrows in Fig. 5.8 indicate the communication pattern between different processors. Little arrows represent communications between four processors as a group working on the same time step. The large arrow represents the communications between two groups of processors working on different time steps, in which each processor in one group communicates with the corresponding processor holding the same spatial subdomain in the other group. It was reported in [189] that the total speedup achieved using space-time parallel algorithm is approximately the product of the speedups obtained by parallelizing the computations for the spatial and the temporal dimensions respectively.

5.4 Window Relaxation

The window relaxation algorithm was first proposed by Saltz et al. in [157, 158]. The purpose of using the window relaxation is to reduce the number of inter-processor communications.

In the traditional sequential time stepping algorithm, a space domain is decomposed into subdomains which are then assigned to different processors. At a particular time step, a processor carries out relaxations for the grid points in its own subdomain. Inter-processor communications are needed after each relaxation to exchange values at the grid points on the boundaries of the subdomains.

As shown by Fig. 5.9, a processor holding a two-dimensional subdomain must exchange solutions on the boundaries of the subdomain with adjacent subdomains. This is done for each iteration i at each time step t_k.

One way to reduce the number of inter-processor communications is to let each processor iterate over a few time steps before making any inter-processor communications. The time steps involved in the relaxations between interprocessor communications constitute a window. Fig. 5.10 shows the window iterations on a processor holding a subdomain Ω_j, where we have assumed that u^0 is the given initial condition and totally $i - 1$ relaxations have been completed. In the ith relaxation, u_i^1 is first calculated, and is then used in the right hand side as an approximation to u^1 for calculating u_i^2. The value u_i^2 is then used in the same way to calculate u_i^3. There are totally three time steps in the window. The boundary values (values at the ghost points) are all from the previous iteration. After a complete iteration which has three stages for the three time steps respectively in the window, the processor needs to

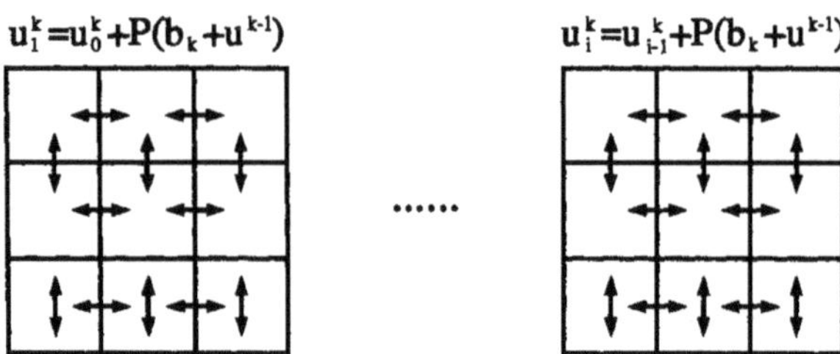

Figure 5.9: In the sequential time stepping algorithm, inter-processor communications indicated by arrows between subdomains are required after each relaxation. The letter k represents the time step and i the iteration index.

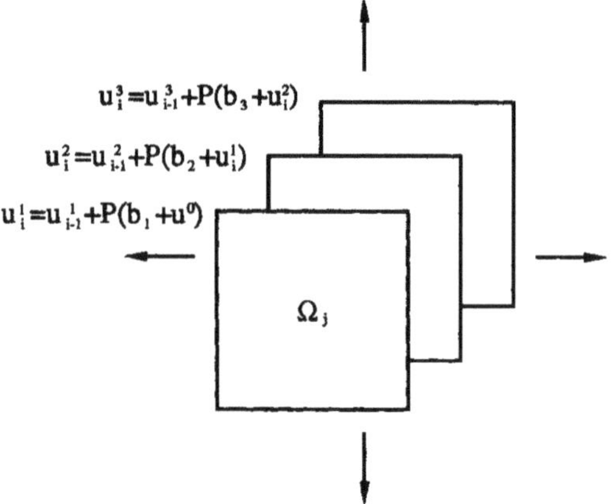

Figure 5.10: In the window relaxation algorithm with a three-time-step window, a processor holding a subdomain Ω_j carrys out relaxations over three time steps, with one relaxation at each time step, before exchange a message with the neighboring processors. The arrows represent interprocessor communications.

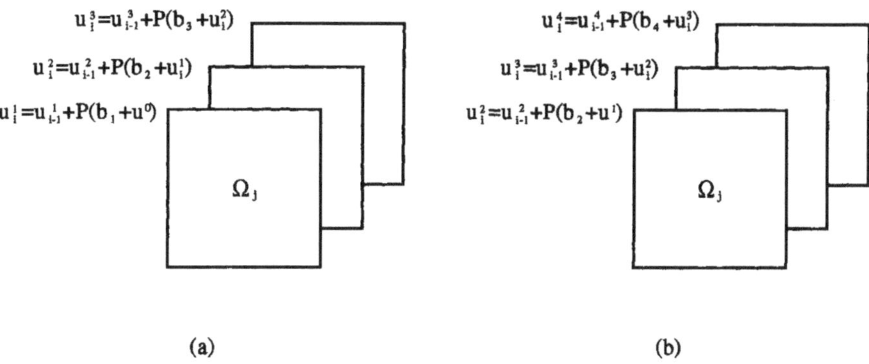

Figure 5.11: (a) The initial window includes time steps 1, 2, and 3 on a processor holding subdomain Ω_j. (b) The window is shifted forward to include time steps 2, 3, and 4 after the relaxation has converged at time step t_1.

exchange the values just calculated on the boundaries of the subdomain with the processors holding adjacent subdomains. Note that the message length of each message exchange is tripled since three time steps are involved. However, if the relaxations over the three-time-step window can converge in about the same number of iterations as required by the relaxations on a single time step, then the total number of inter-processor communications in the window relaxation is only one third of that in the single step relaxation, since there is only one inter-processor communication for ever three time steps in the window relaxations. In this case, the aggregate amount of data exchanged in both algorithms are about the same. With the window relaxation algorithm, there are fewer message exchanges with a longer message length as compared to the single step relaxation. Since on most distributed memory parallel computers, like the Intel iPSC/860 and the Paragon, the communication start-up time (the time required to get the communication channels ready to send or receive a message) is usually higher than the time for sending a data by several orders of magnitude, it is more efficient to send a few large messages than to send many small messages. Therefore, the window relaxation algorithm can save communication time by reducing the start-up time.

In practical computations, the window relaxation is likely to converge for the first time step before the convergence criterion is satisfied for all the other time steps in the window. In this case, it is no longer necessary to keep the first time step in the window. To save the total amount of numerical computations, the window should be shifted in the time dimension to include a new time step in the relaxations.

Fig. 5.11, shows the progress of the window relaxation with a window size of three. The relaxation starts with the initial window including time steps 1, 2 and 3. After certain number of iterations, the solution u_i^1 should be the first, among u_i^1, u_i^2 and u_i^3, to satisfy the convergence criterion at time step t_1. The window will then shift one time step forward, as shown by Fig. 5.11(b), to include time step t_4 in the relaxation. The process continues until the final time step t_m is included in the window. After that, the window size will gradually decrease as the time steps at which the convergence criteria have been satisfied leave the window, until the relaxation has converged at the final time step t_m. The algorithm of the window relaxation can be summarized as:

$$
\boxed{
\begin{array}{l}
\textbf{Algorithm 5.5} \\
\textbf{Window Relaxation Algorithm} \\
\\
u^0 = \text{initial condition} \\
\textbf{do } k = 1, m \\
\quad u_0^k = u^{k-1} \\
\quad i = 0 \\
\quad \textbf{do until } (converge\ at\ step\ t_k) \\
\quad\quad i = i + 1 \\
\quad\quad \textbf{do } j = 0, \min(size - 1, m - k) \\
\quad\quad\quad u_i^{k+j} = u_{i-1}^{k+j} + P(b_{k+j} + u_i^{k+j-1}) \\
\quad\quad \textbf{end do} \\
\quad \textbf{end do} \\
\quad u^k = u_i^k \\
\textbf{end do}
\end{array}
}
$$

where m is the number of time steps and $size$ is the window size. The reason for the upper bound of the inner most do-loop to be $\min(size - 1, m - k)$ is because there may not be enough time steps left to fill up the window toward the end of the solution process. Also the relaxation in the algorithm can be in a more general form as

$$
u_i^{k+j} = \phi(u_{i-1}^{k+j}, u_i^{k+j-1}).
$$

At this point, one may have noticed that the reduction of the communication time using the window relaxation algorithm is based on the assumption that the relaxations for all the time steps in a window will converge in about the same number of iterations as required by the relaxations for a single time step. While it is true that for linear PDEs the asymptotic convergence rates of the Jacobi and Gauss-Seidel type algorithms are the same for both the window relaxation and the single step relaxation (to be shown later in this chapter), the numbers of iterations required to satisfy a given convergence criterion are different for the window and the single step relaxations. For the example given in Fig. 5.11, assume that it will take 10 iterations for the relaxation to converge at each single time step, then totally 30 iterations are needed for computing the solutions at three time steps using the single step relaxation. With window relaxations, however, the relaxations will converge after 10 iterations only for the solution u_i^1 at time step t_1. The solutions u_i^2 and u_i^3 are not converged since the previous ten relaxations at time step 2 and 3 are based on the non-converged values of u_i^1 and u_i^2 respectively in the right hand side of the equations. If two more iterations are required for each of u_i^2 and u_i^3 to converge, as shown in Fig. 5.12, then the total amount of computations for calculating solutions u_i^1, u_i^2 and u_i^3 at time step 1, 2 and 3 respectively is equivalent to 36 single step iterations, which means an increase of 20% in numerical computations as compared with the single step relaxations. As more time steps are used in a window, it is clear from this analysis that the total amount of numerical computations required to compute solutions at all time steps in a window will also increase as well. If the reduction in communication time is more than what is needed to offset the increase in the computation time, then the window relaxation algorithm will be more efficient than the single step relaxation algorithm. As can be seen from these discussions, increasing the window size reduces the time for inter-processor communications, but requires more numerical computations to calculate the solutions. The optimal window size depends on the problem to be solved (the PDE), the relaxation algorithm (Jacobi, Gauss-Seidel, or SOR), and the computer architecture (the ratio of the computation speed to the communication speed). It is shown in [158] that for solving the heat equation on a unit square with Dirichlet boundary conditions, the window relaxation algorithm with a window size of 2 or 3 can reduce the total run time significantly. The algorithm was based on the block SOR iterations and was implemented on Intel iPSC/1 hypercubes.

It should be noted that the execution of the window relaxation algorithm on each processor is actually sequential in the time dimension, i.e. the iterations proceed from the first time step sequentially to the last time step in a window, as shown in Fig. 5.10. The algorithm can be considered as a parallel time stepping algorithm in the following senses:

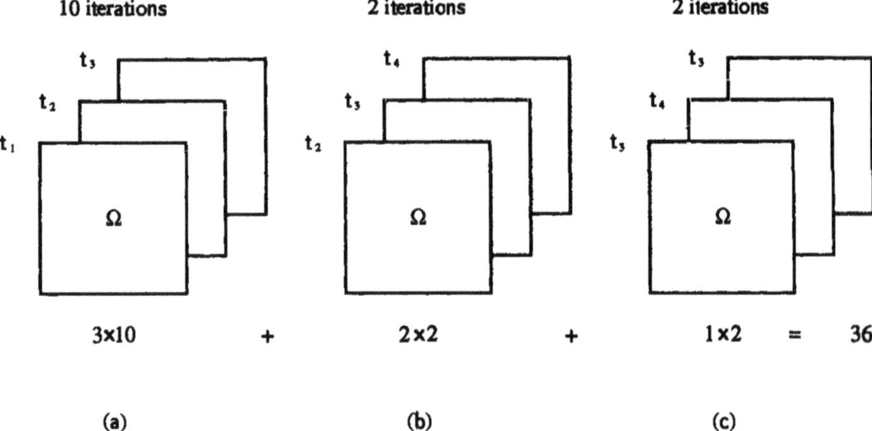

Figure 5.12: (a) Ten iterations at all three time steps t_1, t_2 and t_3 in the initial window. The amount of computations is equivalent to 30 iterations of the single step relaxations. (b) Two more iterations at time steps t_2, t_3 and t_4 in the shifted window. The total amount of computations for time steps t_2 and t_3 is equivalent to 4 iterations of the single step relaxations. (c) Two more iterations at time steps t_3, t_4 and t_5. The total amount of computations for time step t_3 is equivalent to 2 iterations of the single step relaxations.

- On a particular processor, the relaxation proceeds to a later time step (although sequentially) before the solutions at previous time steps have converged. Therefore, there are several time steps actively participating the window relaxation as opposed to the single step relaxation in which only one time step is active in the iterations.

- Different processors may work on different time steps in a window. Assume that there are two processors and the window size is 3. As shown in Fig. 5.13, if for

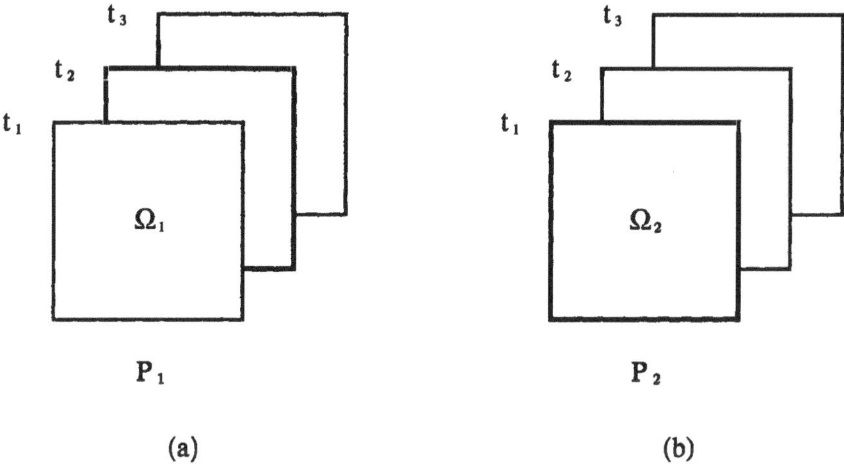

Figure 5.13: (a) Processor P_1 works on the second time step represented by the highlighted square. (b) Processor P_2 still works on the first time step represented by the highlighted square.

some reason that processor P_1 has finished the iteration at time step t_1 before processor P_2 does, it can proceed to the iteration at time step t_2 while processor P_2 still works on the iteration at time step t_1. This can happen when subdomain Ω_1 is smaller than subdomain Ω_2 so that there is less computations to do in Ω_1, or when processor P_1 finished inter-processor communications earlier than does processor P_2 since inter-processor communications can not be synchronized perfectly in general as discussed in Sections 3.2 and 3.4, or when the overall system load is heavier on processor P_2 than that on processor P_1 if processors are shared by more than one users, just name a few possible situations. There

is an inter-processor communication between the two processors in the end of
a complete window relaxation (meaning one iteration for each time step in
the window), so if processor P_1 gets to the end of the window faster than P_2
does, it waits for P_2 to exchange the solution values on the boundaries of the
subdomains.

5.5 Parabolic Multigrid Method

The parabolic multigrid method was first proposed by Hackbusch [74], and later
analyzed more comprehensively in [15, 23], and applied to the unsteady heat equation
in [93, 109] and the Navier-Stokes equations in [92]. The algorithm was developed
based on the direct PDE formulation and can be described by the following procedures
in which we use only two levels of grids and the V-cycle multigrid algorithm for
simplicity of the discussion:

Algorithm 5.6
Parabolic Multigrid Algorithm

Initialization

do $i = 0$ until converge

    do $k = 1, m$ in parallel

        1. pre-smoothing ν_1 times $u^h_{k,i+\frac{1}{3}} = S^h(u^h_{k,i}, u^h_{k-1,I})$

        2. calculating residual of Eq. (5.2)
$$r^h_k = A^h u^h_{k,i+\frac{1}{3}} - b_k - C u^h_{k-1,i+\frac{1}{3}}$$

        3. projecting to coarse grid
$$r^{2h}_k = I^{2h}_h r^h_k$$

        4. solving coarse grid equation
$$A^{2h} \Delta u^{2h}_k = r^{2h}_k + D \Delta u^{2h}_{k-1}$$

        5. correcting solution $u^h_{k,i+\frac{2}{3}} = u^h_{k,i+\frac{1}{3}} + I^h_{2h} \Delta u^{2h}_k$

        6. post-smoothing ν_2 times $u^h_{k,i+1} = S^h(u^h_{k,i+\frac{2}{3}}, u^h_{k-1,i+\frac{2}{3}})$

    end do

    $i = i + 1$

end do

In the above algorithm, h represents the grid spacing, k represents the time step,
and i represents the number of V-cycle iterations. We assume that the equation at

time t_k is assigned to processor P_k. The operation for all m time steps are carried out in parallel. However, there are many inter-processor communications involved at each step in one V-cycle multigrid iteration. In the first step of the algorithm, if the index I is set as i, then there will be no inter-processor communication in the ν_1 smoothing operation. If $I = i + \frac{1}{3}$, the processor computing $u^h_{k-1,i+\frac{1}{3}}$ must send it to the processor computing $u^h_{k,i+\frac{1}{3}}$ after each of the ν_1 smoothing operations. In either case, at the end of the ν_1 smoothing operations, processor P_{k-1} must send the computed $u^h_{k-1,i+\frac{1}{3}}$ to processor P_k for computing the residual r^h_k. Similarly, processor P_k must send $u^h_{k,i+\frac{1}{3}}$ to processor P_{k+1} for computing the residual r^h_{k+1}. At the end of step 4, processor P_k needs to send the computed Δu^{2h}_k to processor P_{k+1} and receives Δu^{2h}_{k-1} from processor P_{k-1} to prepare for the computations in the next V-cycle. Finally, at the end of step 5 and step 6, similar communication procedures are needed to pass $u^h_{k,i+\frac{2}{3}}$ and $u^h_{k,i+1}$ respectively between different processors for the computation of the next V-cycle.

The convergence rate of the multigrid method applied simultaneously to many time steps stays fairly stable as compared with that of the single time step case. This is mainly due to the fact that there are frequent inter-processor communications between different steps in a V-cycle iteration, so that the latest computed information at different time steps propagates efficiently along the time dimension.

For more than two levels of grids , step 4 of the Parabolic Multigrid Algorithm can be modified so that the coarse grid equation is solved using the multigrid algorithm recursively. The algorithm can also be extended to use the W-cycle or the full multigrid algorithms. For details, readers are referred to [15, 23, 74, 92, 94, 93, 109].

5.6　General Form of Parallel Time Stepping Algorithms

Several parallel time stepping algorithms based on either the ODE or the PDE formulation have been discussed in the previous sections. We will show in this section that all those algorithms can be put into a unified form. To simplify the discussion, we will use the one dimensional heat equation

$$\frac{\partial u}{\partial t} = \frac{\partial^2 u}{\partial x^2} + q(x,t) \qquad (x,t) \in [0,1] \times [0,T]$$

$$u(0,t) = f(t), \qquad u(1,t) = g(t) \ \ t > 0 \qquad (5.21)$$

$$u(x,0) = h(x) \qquad x \in (0,1)$$

to demonstrate how the general algorithm is formulated and different special algorithms are obtained.

Assume that the space domain is discretized by N grid points x_i, $i = 1, 2, \cdots N$, and the time interval is represented by m steps t_k, $k = 1, 2, \cdots m$. Discretizing Eq. (5.21) using the first order backward time differencing and the second order central space differencing for the time and the space derivatives respectively, we will have at each grid point (x_i, t_k)

$$\frac{u_i^k - u_i^{k-1}}{\Delta t} = \frac{u_{i-1}^k - 2u_i^k + u_{i+1}^k}{\Delta x^2} + q_i^k \tag{5.22}$$

$$i = 1, 2, \cdots N, \qquad k = 1, 2, \cdots m.$$

Eq. (5.22) can be re-written as

$$-\lambda u_{i-1}^k + (1 + 2\lambda)u_i^k - \lambda u_{i+1}^k = u_i^{k-1} + \Delta t q_i^k \tag{5.23}$$

$$i = 1, 2, \cdots N, \qquad k = 1, 2, \cdots m,$$

where $\lambda = \Delta t / \Delta x^2$.

Assembling Eq. (5.23) at all spatial grid points x_i, $i = 1, 2, \cdots N$, we obtain a matrix equation system

$$Au^k = u^{k-1} + q^k, \quad k = 1, 2, \cdots m. \tag{5.24}$$

with

$$A = \begin{bmatrix} (1+2\lambda) & -\lambda & & \\ -\lambda & (1+2\lambda) & \ddots & \\ & \ddots & \ddots & -\lambda \\ & & -\lambda & (1+2\lambda) \end{bmatrix}, \tag{5.25}$$

$$u^k = \begin{bmatrix} u_1^k \\ u_2^k \\ \vdots \\ u_N^k \end{bmatrix}, \quad q^k = \begin{bmatrix} \Delta t q_1^k + \lambda f^k \\ \Delta t q_2^k \\ \vdots \\ \Delta t q_{N-1}^k \\ \Delta t q_N^k + \lambda g^k \end{bmatrix}.$$

The traditional sequential time stepping algorithm starts with the first equation system ($k = 1$) in (5.24) and solves all equation systems one by one. The first equation is solved using the initial condition

$$u^0 = (h(x_1), h(x_2), \cdots, h(x_N))^T,$$

and the solution u^k is put into the right hand side of the equation system at time step t_{k+1}.

For parallel time stepping algorithms, we want processors to work on multiple time steps simultaneously. To this end, we can further assemble all equation systems in Eq. (5.24) with different index k into an augmented system, just like what we did in going from Eq. (5.23) to Eq. (5.24)

$$
\begin{bmatrix}
A & & & \\
-I & A & & \\
& \ddots & \ddots & \\
& & -I & A
\end{bmatrix}
\begin{bmatrix}
u^1 \\ u^2 \\ \vdots \\ u^{m-1} \\ u^m
\end{bmatrix}
=
\begin{bmatrix}
u^0 + q^1 \\ q^2 \\ \vdots \\ q^{m-1} \\ q^m
\end{bmatrix}
\tag{5.26}
$$

where I is the identity matrix.

Eq. (5.26) can be solved using many different parallel algorithms, thus exploiting parallelism in both the temporal and the spatial dimensions. The parallel time stepping algorithms discussed before can all be derived by applying a particular solution algorithm to Eq. (5.26).

1. Waveform Relaxation Algorithm

To get the waveform relaxation algorithm discussed in Section 5.2, we need to re-order all the equations in Eq. (5.26). Instead of numbering the values of u at space grid points first, followed by the time steps, we need to number the values of u at different time steps t_k first, followed by the space grid points. If we assemble Eq. (5.23) by index k first, the matrix equation at each grid point will look like

$$
-\lambda u_{i-1} + B u_i - \lambda u_{i+1} = q_i
\tag{5.27}
$$

$$
i = 1, 2, \cdots N
$$

where B is in the form of

$$
B =
\begin{bmatrix}
(1 + 2\lambda) & & & \\
-1 & (1 + 2\lambda) & & \\
& \ddots & \ddots & \\
& & -1 & (1 + 2\lambda)
\end{bmatrix},
$$

and

$$
q_i = \left(\Delta t q_i^1 + h_i, \Delta t q_i^2, \cdots, \Delta t q_i^m \right)^T,
$$
$$
u_i = \left(u_i^1, u_i^2, \cdots, u_i^m \right)^T.
$$

Now if we further assemble all equation systems in Eq. (5.27) with respect to index i, the final augmented system will be

$$\begin{bmatrix} B & -\lambda I & \ddots & \\ -\lambda I & B & & \\ & \ddots & \ddots & -\lambda I \\ & & -\lambda I & B \end{bmatrix} \begin{bmatrix} u_1 \\ u_2 \\ \vdots \\ u_N \end{bmatrix} = \begin{bmatrix} \Delta t q_1^k + \lambda u_0 \\ \Delta t q_2^k \\ \vdots \\ \Delta t q_N^k + \lambda u_{N+1} \end{bmatrix} \tag{5.28}$$

where $u_0 = (f(t_1), f(t_2), \cdots, f(t_m))^T$ and $u_{N+1} = (g(t_1), g(t_2), \cdots, g(t_m))^T$ are calculated from the boundary conditions given in Eq. (5.21). If we apply the block Jacobi relaxations to Eq. (5.28), we will have

$$Bu_i^{(j+1)} = \lambda u_{i+1}^{(j)} + \lambda u_{i-1}^{(j)} + \Delta t q_i \tag{5.29}$$

$$i = 1, 2, \cdots N, \qquad j = 0, 1, \cdots,$$

which is just the formula for the Jacobi waveform relaxation as shown by Eq. (5.8). The Gauss-Seidel and SOR waveform relaxation can be formulated similarly. If the solution process of each of the bi-diagonal systems in Eq. (5.29) is also parallelized, we then have the space and time parallel waveform relaxation algorithm discussed in Section 5.2.3.

Note that Eq. (5.28) can also be obtained directly from Eq. (5.26) by exchanging proper rows and columns in the augmented matrix, and reordering the unknowns in the index k first, followed by the index i.

2. Pipeline Iteration

To obtain the pipeline iteration algorithm discussed in Section 5.3, we can apply any relaxation algorithm directly to the augmented matrix equation system (5.26). For example, the block Gauss-Seidel iteration for Eq. (5.26) can be represented as

$$\begin{bmatrix} -L+D & & & \\ -I & L+D & & \\ & \ddots & \ddots & \\ & & -I & L+D \end{bmatrix} \begin{bmatrix} u^1 \\ u^2 \\ \vdots \\ u^m \end{bmatrix}_i = \tag{5.30}$$

$$\begin{bmatrix} U & & & \\ & U & & \\ & & \ddots & \\ & & & U \end{bmatrix} \begin{bmatrix} u^1 \\ u^2 \\ \vdots \\ u^m \end{bmatrix}_{i-1} + \begin{bmatrix} g^1 + u^0 \\ q_2 \\ \vdots \\ q^m \end{bmatrix},$$

$$i = 1, 2, \cdots,$$

where we have assumed that matrix A has been split into three parts: the lower triangular part $-L$, the diagonal part D, and the upper triangular part $-U$. Eq. (5.30) can also be written in a block form as

$$u_i^k = (L + D)^{-1}[Uu_{i-1}^k + u_i^{k-1} + q^k] \tag{5.31}$$

$$k = 1, 2, \cdots m, \quad i = 1, 2, \cdots,$$

where $u_i^0 = u^0$ is the initial condition. We see from Eq. (5.31) that for a given Gauss-Seidel iteration (a particular i), the computation starts from the equation at the first time step ($k = 1$), the computed value u_i^1 is used in the right hand side of the equation at the second time step ($k = 2$). To accommodate more general type of relaxation algorithms, Eq. (5.31) can be written as

$$u_i^k = \phi(u_{i-1}^k, u_i^{k-1}) \tag{5.32}$$

$$k = 1, 2, \cdots m, \quad i = 1, 2, \cdots,$$

which is exactly the formula used in the pipeline iteration algorithm discussed on page 191.

3. Window Relaxation

If we decompose the domain $[0, 1]$ into several subdomains, say three identical subdomains, and write Eq. (5.22) in block form corresponding to the three subdomains, we will have

$$Au^k = u^{k-1} + q^k \quad k = 1, 2, \cdots m \tag{5.33}$$

where

$$A = \begin{bmatrix} C & E & \\ E^T & C & E \\ & E^T & C \end{bmatrix}, \quad u^k = \begin{bmatrix} u_1^k \\ u_2^k \\ u_3^k \end{bmatrix}. \tag{5.34}$$

In Eq. (5.34), u_i^k represents the unknown vector in the i-th subdomain at the k-th time step, matrix C has the same structure as that of A in (5.25), and all elements in matrix E are zero, except the element $-\lambda$ in the lower left corner. Using three time steps, we have the augmented matrix equation system as

$$\begin{bmatrix} A & & \\ -I & A & \\ & -I & A \end{bmatrix} \begin{bmatrix} u^1 \\ u^2 \\ u^3 \end{bmatrix} = \begin{bmatrix} q^1 + u^0 \\ q^2 \\ q^3 \end{bmatrix}. \tag{5.35}$$

Now let us re-order the unknowns in Eq. (5.35) as

$$
\begin{bmatrix} u^1 \\ u^2 \\ u^3 \end{bmatrix} =
\begin{bmatrix} u_1^1 \\ u_2^1 \\ u_3^1 \\ u_1^2 \\ u_2^2 \\ u_3^2 \\ u_1^3 \\ u_2^3 \\ u_3^3 \end{bmatrix}
\longrightarrow
\begin{bmatrix} u_1^1 \\ u_1^2 \\ u_1^3 \\ u_2^1 \\ u_2^2 \\ u_2^3 \\ u_3^1 \\ u_3^2 \\ u_3^3 \end{bmatrix} =
\begin{bmatrix} u_1 \\ u_2 \\ u_3 \end{bmatrix}
\tag{5.36}
$$

where we have changed ordering from space-dimension-first to time-dimension-first. The rows and columns in the coefficient matrix of Eq. (5.35) should also be exchanged correspondingly, so that the final equation system becomes

$$
\begin{bmatrix} B & F & \\ F^T & B & F \\ & F^T & B \end{bmatrix}
\begin{bmatrix} u_1 \\ u_2 \\ u_3 \end{bmatrix} =
\begin{bmatrix} q_1 \\ q_2 \\ q_3 \end{bmatrix}
\tag{5.37}
$$

where

$$
B = \begin{bmatrix} C & & \\ -I & C & \\ & -I & C \end{bmatrix}, \qquad
F = \begin{bmatrix} E & & \\ & E & \\ & & E \end{bmatrix},
$$

and all components in u_i and q_i are re-ordered from u^i and q^i in Eq. (5.35). Note that u_i represents unknowns at three time steps in the i-th subdomain, while u^i represents unknowns in all three subdomains at the i-th time step. Now if we apply the block Jacobi relaxations to Eq. (5.37), the iterations can be written as

$$
\begin{aligned}
Bu_1^{(j)} &= q_1 - Fu_2^{(j-1)} \\
Bu_2^{(j)} &= q_2 - F^T u_1^{(j-1)} - Fu_3^{(j-1)} \\
Bu_3^{(j)} &= q_3 - F^T u_2^{(j-1)} \\
j &= 1, 2 \cdots .
\end{aligned}
\tag{5.38}
$$

In each j-iteration, the three block equations in Eq. (5.38) can be solved independently on three different processors since the right hand side terms depend only on the values from the previous j-iteration. From Eq. (5.37), we see that each of the three equations in Eq. (5.38) includes unknown vectors at three time steps in the i-th subdomain. Therefore, the iterations represented by Eq. (5.38) is just the window iteration with a window size of three. Depending on how the unknown vector

$u_i^{(j)}$, $i = 1, 2, 3$, is solved from Eq. (5.38) in each j-iteration, we can have different window relaxation algorithms (like Jacobi, Gauss-Seidel and SOR window iterations) discussed in Section 5.4.

4. Parabolic Multigrid

The parabolic multigrid algorithm discussed in Section 5.5 can be considered as the direct application of the multigrid algorithm to Eq. (5.26), with the exception that there is no grid coarsening in the time dimension. It has been found that since the propagation of information in the time dimension is usually fundamentally different from that in the spatial dimension for most PDE applications, grid coarsening in the time dimension does not improve the convergence rate significantly. Sometimes it can even cause stability problems.

5.7 Complexity Analysis

It has been discussed in the previous section that parallel time stepping algorithms for solving

$$Au^k = u^{k-1} + b^k, \quad k = 1, 2, \cdots m, \tag{5.39}$$

where k represents time steps, are essentially the applications of different solution algorithms to the augmented system

$$Gu = \begin{bmatrix} A & & & \\ -I & A & & \\ & \ddots & \ddots & \\ & & -I & A \end{bmatrix} \begin{bmatrix} u^1 \\ u^2 \\ \vdots \\ u^{m-1} \\ u^m \end{bmatrix} = \begin{bmatrix} b^1 + u^0 \\ b^2 \\ \vdots \\ b^{m-1} \\ b^m \end{bmatrix}. \tag{5.40}$$

The efficiency of the parallel time stepping algorithms is therefore dependent on how fast the iterations for Eq. (5.40) converge as compared to that for Eq. (5.39) based on the same iterative algorithm. Since the amount of computations in one iteration for Eq. (5.40) is about m times as that in one iteration for Eq. (5.39), the iteration process for Eq. (5.40) has to converge in the number of iterations close to that needed by the iteration for Eq. (5.39) in order to maintain the same numerical complexity. Otherwise, the gain from exploiting parallelism in the temporal dimension can very well be reduced by the increased numerical complexity (the extra amount of computations needed for iterating directly on the augmented system).

Asymptotically, we can show that the convergence rates for both Eq. (5.39) and Eq. (5.40) are the same if the same relaxation algorithm is used. This is because the spectral distribution of the block bi-diagonal matrix G in (5.40) is the same as that of the block matrices in the main diagonal, which is actually matrix A in (5.39). Therefore, both matrices have the same eigenvalues (although with different multiplicity) which determine the convergence rates of most relaxation algorithms.

To justify this conclusion for basic relaxation type algorithms, let us consider Eq. (5.39) and a corresponding iterative algorithm

$$u_i^k = Tu_{i-1}^k + D(b_k + u^{k-1}), \qquad i = 1, 2, \cdots, \tag{5.41}$$

for solving Eq. (5.39). It has been proven that the convergence rate of (5.41) depends on $\rho(T)$, the spectral radius of T [142]. The smaller the $\rho(T)$, the faster the iterations converge. According to the definition, we have

$$\rho(T) = max|\lambda_i|, \qquad i = 1, 2, \cdots, n,$$

where λ_i's are the eigenvalues of the iterative matrix T.

For the Gauss-Seidel iterations, we now show that the iterative matrices for Eq. (5.39) and (5.40) have the same eigenvalues, and hence the same spectral radii.

The iterative matrix T_A of the Gauss-Seidel relaxation algorithm for Eq. (5.39) is

$$T_A = (D - L)^{-1}U$$

where the coefficient matrix A is split as $A = D - L - U$ with D, $-L$ and $-U$ the diagonal, the lower and the upper triangular parts of A respectively. Correspondingly, the iterative matrix $\mathbf{T_G}$ for Eq. (5.40) is

$$\mathbf{T_G} = (\mathbf{D} - \mathbf{L})^{-1}\mathbf{U}$$

where

$$\mathbf{D} = \begin{bmatrix} D & & & \\ & D & & \\ & & \ddots & \\ & & & D \end{bmatrix}, \qquad -\mathbf{L} = \begin{bmatrix} -L & & & \\ -I & -L & & \\ & \ddots & \ddots & \\ & & -I & -L \end{bmatrix},$$

$$-\mathbf{U} = \begin{bmatrix} -U & & & \\ & -U & & \\ & & \ddots & \\ & & & -U \end{bmatrix}.$$

It is easy to verify that

$$(\mathbf{D} - \mathbf{L})^{-1} = \begin{bmatrix} D-L & & & \\ -I & D-L & & \\ & \ddots & \ddots & \\ & & -I & D-L \end{bmatrix}^{-1}$$

$$
= \begin{bmatrix} (D-L)^{-1} & & & \\ (D-L)^{-2} & (D-L)^{-1} & & \\ \vdots & & \ddots & \ddots & \\ (D-L)^{-n} & \cdots & & (D-L)^{-2} & (D-L)^{-1} \end{bmatrix},
$$

so that

$$
\mathbf{T_G} = (\mathbf{D}-\mathbf{L})^{-1}\mathbf{U} = \begin{bmatrix} (D-L)^{-1}U & & & \\ (D-L)^{-2}U & (D-L)^{-1}U & & \\ \vdots & & \ddots & \ddots & \\ (D-L)^{-n}U & \cdots & & (D-L)^{-2}U & (D-L)^{-1}U \end{bmatrix}.
$$

Since $\mathbf{T_G}$ is a block lower triangular matrix, the eigenvalues of $\mathbf{T_G}$ are the collections of all eigenvalues of all block matrices in the main diagonal. It is obvious that all blocks in the main diagonal of $\mathbf{T_G}$ are the same matrix $T_A = (D-L)^{-1}U$, hence the iterative matrix $\mathbf{T_G}$ for Eq. (5.40) has the same eigenvalues, with higher multiplicity, as that of the iterative matrix T_A for Eq. (5.39). This proves

$$
\rho(\mathbf{T_G}) = \rho(T_A)
$$

Therefore, the convergence rates of the Gauss-Seidel iterations are the same for both Eqs. (5.39) and (5.40).

Similar conclusions can be obtained for the Jacobi and SOR iterations in the same way.

Although the asymptotic convergence rates are the same for both Eqs. (5.39) and (5.40), the actual numbers of iterations needed for convergence could be significantly different for the two systems. There are two important factors that cause this difference.

1. Poorer initial guess. When the sequential time stepping algorithm is used to solve Eq. (5.39) at different time steps sequentially, the given initial condition u^0 can be used as a good initial guess of u^1 to solve the first equation in Eq. (5.39)

$$
Au^1 = u^0 + b^1
$$

iteratively. After u^1 has been solved, it can be used again as an initial guess for u^2 to start the iteration at time step t_2. Therefore, with sequential time stepping, we can say that for each equation system defined at a particular time step, the initial guess used to start the iterative solution process is no more than one time step away from the right solution at that time step. However, with parallel time stepping algorithm, the solution process for the augmented system needs initial guesses for the solutions at all time steps to begin the iteration. Therefore, the initial guesses for the later time steps used in the computation could be much farther away from the right solution than in the sequential algorithm. As a result of using poorer initial guesses, the norm of the residual vector will also increase as more time steps are used simultaneously in the parallel time stepping algorithm, as shown in Fig. 5.14.

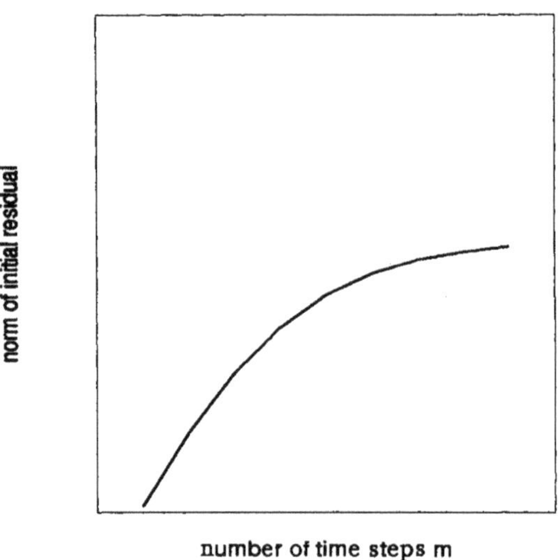

Figure 5.14: Norm of the initial residual vector of the augmented equation system (5.40). The norm increases as the number of time steps increases.

2. Slower convergence rate. In general, the iterative solution process for solving Eq. (5.39) is accelerated by many different schemes, like the multigrid method, as discussed in previous sections. Therefore, the actual rates of convergence observed in those accelerated solution algorithms are much better than the asymptotic convergence rate represented by the spectral radius of the iteration matrix. As more time steps are used simultaneously in the parallel time stepping algorithm, the observed convergence rate of the augmented equation system Eq. (5.40) is approaching the asymptotic convergence rate, as shown in Fig. 5.15. This degraded convergence rate also causes more iterations for the augmented system Eq. (5.40) to converge.

To control the increasing computational complexity, we should avoid using too many time steps simultaneously in the parallel time stepping algorithm. The parallel time stepping algorithm is best suited for the situation when space domain decomposition alone can not utilize the available processors efficiently. In other words, if

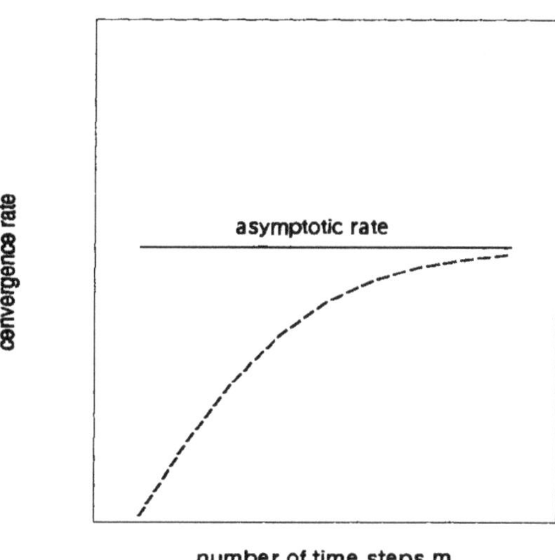

Figure 5.15: Convergence rate of parallel time stepping algorithms applied to Eq. (5.40). As the number of time steps increases, the convergence rate of the relaxation algorithm approaches the asymptotic rate.

the space domain decomposition alone results in very small subdomains, then the parallel time stepping algorithm should be used to increase the amount of numerical computations between message exchanges.

Fig. 5.16 shows the timing curves [200] of the parallel and sequential time step-

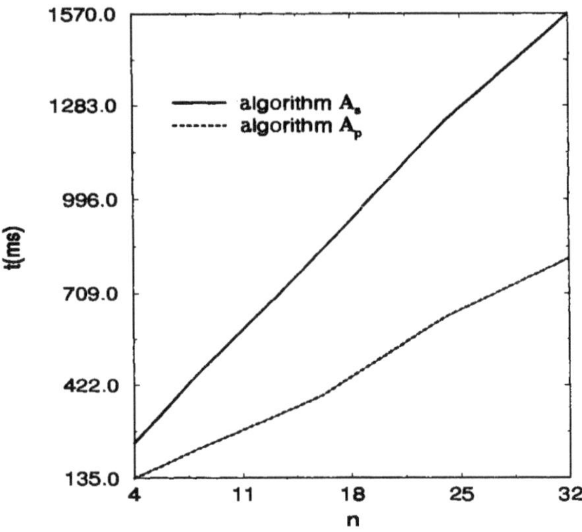

Figure 5.16: Timing curves for the sequential (A_s) and the parallel (A_p) time stepping algorithms.

ping algorithms. In the figure, A_s represents the single step relaxation applied to Eq. (5.39), and A_p represents the multi-step relaxation applied to Eq. (5.40). The following model equation was used to study the performance of algorithms A_s and A_p:

$$\frac{\partial u}{\partial t} + u = \Delta u \qquad (x, y) \in \Omega \qquad 0 < t < T \tag{5.42}$$

$$u(x, y, t) = 0 \qquad (x, y) \in \partial\Omega \qquad 0 < t < T$$

$$u(x, y, 0) = sin(\frac{\pi}{l}x)sin(\frac{\pi}{l}y) \qquad (x, y) \in \Omega$$

where Ω is a two-dimensional square domain of dimensions $l \times l$ which is discretized as a 32×32 grid and $\partial\Omega$ is the boundary. The equation was solved on an iPSC/860 hypercube with 32 nodes. The Jacobi iterative algorithm was used as the linear equation solver. The curves in Fig. 5.16 represent the timing results, in milliseconds, for both algorithms as functions of the number of time steps n. It can be seen that the parallel time stepping algorithm is faster than the sequential time stepping algorithm. This is due to the fact that the space grid is not big enough to use 32 processors efficiently. The use of the parallel time stepping algorithm alleviated this problem by increasing the amount of numerical computations between message exchanges.

Chapter

6

Future Development

In this chapter, we will discuss some of the latest developments in parallel computing. These developments are likely to set the directions for the research in this area for the next few years. Since parallel computing is a fast-developing area, we can not expect that the whole picture of the future be depicted completely based on the current available information. The best we can do is to make some predictions based on what has happened. We have seen that over the last two decades new computers emerged every two or three years, while the development in software has somehow lagged behind. The major difficulty that impedes the wide-spread use of parallel computers by researchers and engineers working on practical scientific applications is the lack of truly user-friendly software environments. Therefore, we expect the development of user-friendly software tools and environments to become a major research area in the near future. Another important development in high performance computing is the formation of Meta-computers or Meta-centers which represent a coherent computing resource, including different parallel computers and workstations with multi-media processing capability, shared through a national or international network. To the users, these resources function as a single tightly coupled meta-computer, even if many of the components are located several thousand miles apart. Again, the key for realizing this meta-computer is the software tools and environments that form the user interface.

6.1　More on Performance Evaluations

Since the purpose of using parallel computers is to reduce the turn-around time (the wall-clock time) needed for computing solutions of complicated problems, users of

parallel computers will naturally want to know how much execution time can be
reduced by using more processors in the computation [91, 106, 114, 191, 197]. The
most straightforward metric for measuring this reduction is the speedup S_p discussed
in Section 1.3.1. It is defined as

$$S_p = \frac{T_s}{T_p} \tag{6.1}$$

where T_s is the sequential execution time and T_p is the parallel execution time on p
processors. If T_s is the execution time obtained by using the best sequential algorithm
(this can be a little controversial because for some problems, there is no consensus as
to which algorithm is the best), then S_p is called the absolute speedup [143]. However,
in many cases, it is difficult to use this metric because it often needs implementation
of two different algorithms. Another widely used speedup definition is to take T_s as T_1
— the execution time of the same parallel algorithm on one processor. This is called
the relative speedup [143]. The application of this definition to the architectures with
a few processors is straightforward and has not caused any major problem. However,
as more processors are used in the computation and the memory hierarchy becomes
more complicated, the speedup metric defined by Eq. (6.1) becomes difficult to use
and often gives misleading results.

- Distributed Memory Parallel Computers. On this type of architectures, the
  size of a problem that can be solved on one processor is limited by the available local memory on that processor (currently less than 64 Mbytes for most
  architectures). As can be seen from definition (6.1), to calculate S_p, we need
  the execution times of the same problem on both one and p processors. When
  the number of processors increases, we can not increase the problem size proportionally if Eq. (6.1) is used to evaluate the speedup, since the problem and
  all its data must be small enough to fit into the local memory of one processor.
  In other words, we have to use problems with a fixed size to evaluate the performance of parallel computers and algorithms with S_p defined by Eq. (6.1).
  As discussed in Section 1.3.3, S_p is bounded by the sequential part of a parallel algorithm if the problem size is fixed [6]. As a result, no architecture and
  algorithm will be scalable with fixed size problem.

  A better speedup — the scaled speedup that scales the problem size with
  the number of processors was later proposed in [72, 73] and in [166, 167, 168].
  Instead of fixing the problem size, the scaled speedup fixes either the execution
  time or the memory size. The fixed-time speedup was first proposed in [73]
  and the memory-bounded speedup in [167]. In the fixed-time speedup, a proper
  time limit t_f (say 60 seconds) is selected and the problem size is scaled based

on the number of processors to meet the time limit. The speedup definition is then

$$S_t = \frac{T(n_p)}{T(n_1)} = \frac{T(n_p)}{t_f} \tag{6.2}$$

where n_1 and n_p are the problem sizes that require t_f units of time on one and p processors resectively, $T(n_p)$ is the execution time on one processor to solve the problem with size n_p that takes p processors t_f units of time to solve. This definition considers the scaling of problem size, but the time limit t_f has to be small enough so that the problem with size n_p still can be fit into and solved on one processor.

The memory-bounded speedup uses a similar idea and interested readers are referred to [166, 167, 168] for details.

- Shared Memory Parallel Computers. On shared memory parallel computers, a problem that can be solved on p processors can also in principle be solved on one processor, since a single processor has access to all the available memory on the computer. Therefore, it appears that definition (6.1) is well defined. Table 6.1 shows the timing results for factorizing a 4096×2048 matrix in the

Table 6.1: Timing results and speedups for factorizing a 4096×2048
matrix using the Householder Transformation
t: time (s)
s: speedup

t&s \ p	1	2	3	4	8	16	32	48	56
t	6669	4309	2627	1451	547.4	227.4	133	98.5	87.9
s	1.00	1.55	2.54	4.60	12.18	29.33	50.14	67.71	75.87

solution of the regularized least squares problem on KSR-1 parallel computers [202]. The code is a core subroutine in solving the inverse problems of PDEs related to oil reservoir simulations [204, 205]. This problem requires more than 32 Mbytes of memory. In the table, p is the number of processors used in the computation, t is the execution time in seconds, and s is the speedup calculated using definition (6.1). We can see from Table 6.1 and Fig. 6.1 that the speedup is excellent. Significant superlinear speedup is demonstrated for certain number of processors. The users may therefore be misled to the conclusion that the KSR-1 is a perfect scalable parallel computer for large problems.

What actually happened when a large problem is solved on a single processor of KSR-1 is that the portion of the data that can not be accommodated in that

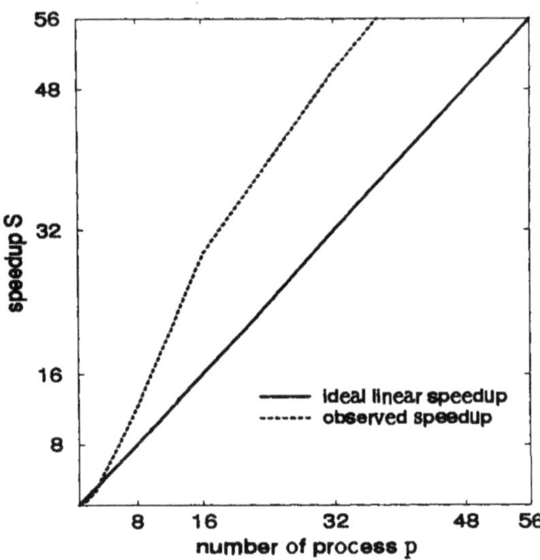

Figure 6.1: Speedup curves for solving the regularized least squares problem on KSR-1. The speedup is calculated using (6.1).

processors local memory is distributed to the local memory of other processors. Since only one processor is doing the computations, all data located in other processors must be brought into the working processor for the computations. As a result, much more data swapping is incurred with only one processor doing the computations than the case when all processors holding the data participate in the computations. Therefore, definition (6.1) is not appropriate for KSR-1 if the problem to be solved requires more memory than what is available on a single processor.

Actually, this difficulty arises for any shared memory parallel computer if there is a memory hierarchy on the architecture such that access to the memory is significantly slower for large problems than for small problems. On the KSR-1, each processor has 32 Mbytes of local memory, if the problem is so big that part of the data must be put into the local memory of other processors, then access to that part of data using different levels of search engines is about 8 times slower. As a result, the computational speed of a single processor will be much slower for a large problem.

The generalized speedup proposed in [166] recognizes the difference in memory access times for a specific problem with different sizes on a single processor. While it is not fair to solve a small problem for evaluating the performance of a parallel computer with many processors, it is inappropriate either to solve a very large problem on a single processor to measure its performance. Therefore, the generalized speedup proposed in [166] defines the speedup as

$$S_g = \frac{\text{parallel speed}}{\text{sequential speed}} \tag{6.3}$$

where speed is defined as the quotient of the work and the elapsed time. The sequential speed is measured by solving a problem with an appropriate size n_1 on one processor, and the parallel speed is measured by solving the same problem with a scaled size n_p on p processors. The detailed discussions of how to choose size n_1 and scale it to n_p for p processors are given in [166].

Table 6.2 shows the speeds for factorizing a $2n \times n$ matrix using the Householder

Table 6.2: Speeds achieved on different numbers of processors
p: number of processors
n: problem size
S_p: speed in MFlops

n & speedup / p	1	2	4	8	16	32	48	56
n	724	1024	1448	2048	2896	4096	5016	5418
S_p	5.4	9.7	20.8	35.1	73.8	134.4	206.4	218.4

transformation on different numbers of processors. This is the same regularized least squares problem reported in Table 6.1. The speed is calculated by dividing the total floating point operations by the elapsed wall-clock time. Using the speeds in Table 6.2 and definition (6.3), the generalized speedup S_g is given in Fig. 6.2 which is more reasonable as compared with Fig. 6.1.

In general, however, the performance evaluations of parallel computers and algorithms are difficult because the selection of initial size n_1 and the way it is scaled to n_p for p processors depend on the architectures and the algorithms being studied. If a practical application problem is used to study the performance of the architecture and the algorithm, it may not be easy or even possible to change the problem size as in the case of matrix computations. For example, if a complete airplane is discretized by a finite element or finite difference scheme, it may take months to define the geometry and generate a grid. Therefore, it

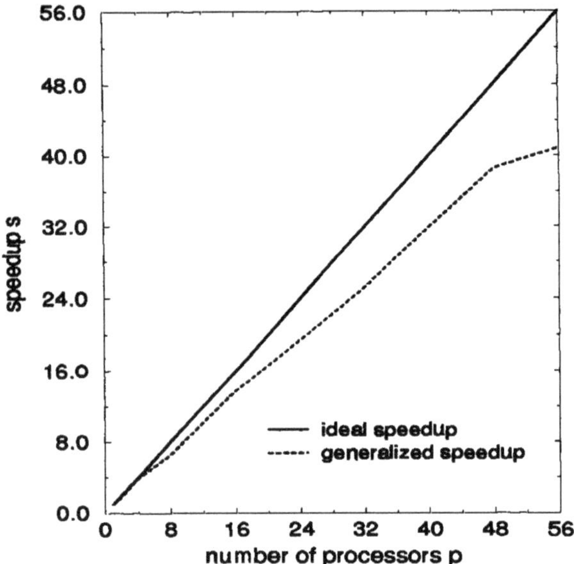

Figure 6.2: Speedup curves for solving the regularized least squares problem of different sizes on different numbers of processors of KSR-1. The speedup is calculated using (6.3).

is impractical to generate many grids with different numbers of grid points for different numbers of processors.

As different new architectures constantly emerge and new applications are being solved on parallel computers, the research in performance evaluation and scalability study will remain an active area for the development of easy-to-use performance metrics.

6.2 Programming Language

Currently, the most widely used programming languages on parallel computers are Fortran and C. Since these languages were not developed originally for parallel computers, different language extensions have been added to these languages by different hardware and software developers. For example, the message passing and synchronization features are added to Fortran and C for Intel hypercube and mesh-structured parallel computers, which is discussed briefly in Section 3.2. Note that on different parallel computers (even if with the same type of architecture like the distributed memory MIMD architecture), language extensions are usually in different forms, although the functions they perform are pretty much the same. As a result of these non-standardized extensions, a user has to modify his existing code, sometimes even change the structure of the code substantially, in order to run the code on a different parallel computer. Given the fact that new parallel computers emerge every two to three years, it is virtually impossible for users running large application codes to constantly modify their codes for different parallel computers.

To overcome this difficulty and make application codes portable across different parallel computers, many researchers have developed programming environments that can support different parallel architectures. The following is an incomplete list of these environments (in Fortran), together with architectures they support. A more comprehensive survey of these environments is given in [36] by Chen.

- HPF (High Performance Fortran) [88], still in the phase of development by a consortium of many universities and vendors, will support architectures from major parallel computer manufactures like Alliant, Convex, Cray Rearch, DEC, Fujitsu, HP, IBM, Intel, MasPar, nCUBE and TMC.

- ADAPTOR (Automatic DAta Parallelism TranslatOR) [18], developed in Germany, supports CM5, KSR-1, iPSC/860 and Alliant Fx/2800.

- P-LANGUAGES [160], developed in University of Houston in USA, supports iPSC/2, iPSC/860, Delta, KSR-1, nCUBE2.

- Force [102, 103], developed at University of Colorado, Boulder, USA, supports Cray Y-MP, Cray 2, KSR-1, Encore, Sequent, Convex, Alliant.

- Vienna Fortran [21, 33, 34], developed at University of Vienna, Austria, supports Intel iPSC/860, SUPRENUM, Genesis-P machine, all distributed memory machines on which PARMACS (v. 5.0) runs.

- Fortran D [54, 89], developed at Rice University, USA, supports Intel iPSC/860, Sun Sparc and IBM RS6000 clusters.

- TOPSYS (TOols for Parallel SYStems) [16, 131], developed at Technical University of Munch, supports iPSC/2, iPSC/860, PARSYTEC SC, EDS.

A common feature of all these environments is that they translate the parallel codes developed in these environments for a particular parallel machine by inserting the machine dependent data distribution, message passing, and synchronization instructions. The translated code is then compiled using the compiler on that particular machine. As shown in Fig. 6.3, the parallel programming environments listed above function just like translators. Since most of those environments support only a small set of available parallel computers, application codes developed in one of this environments is still not truly portable across a wide range of architectures. The only exception is the development of HPF [88] and Fortran 90 [98] which are endorsed by most major parallel computer manufactures. The language specification of Fortran 90 has been published by the International Standardization Organization [1]. The extensions based on Fortran 77 include array operations, user-defined data types, pointers, dynamic memory allocation/deallocation, recursive subroutine calls, and many more other features. The new extensions were proposed to improve Fortran 77 for scientific computing on all platforms, not specifically for parallel computers. The project HPF is aimed at providing standardized computer directives in Fortran 90 for parallel processing, including data distribution and synchronization. Implementation of these extensions on different parallel computers will provide users a unified and machine independent interface for parallel processing, so that application codes developed in this environment can run without major modifications on a wide spectrum of platforms, ranging from desktop workstations to massively parallel computers.

6.3 Automatic Parallelization Tools

The programming languages discussed in Section 6.2 provide users a portable parallel program environment. However, users who develop parallel codes still need to express,

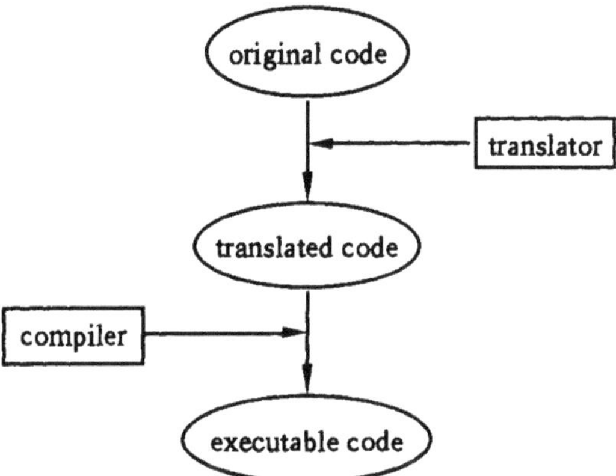

Figure 6.3: Flow chart of translating and compiling a portable program for a parallel computer. The translator inserts the machine specific instructions for message passing, data distribution, and synchronization into the original code, and the compiler on a specific parallel computer generates the executable code.

in one way or the other, parallelism in their algorithms explicitly. This could be a
problem to many researchers and engineers who need the computing power of parallel
computers but are not familiar with parallel algorithms. This situation is somewhat
similar to the difficulty in the early days when people have to program a digital
computer using assembly language or even machine instructions. For example, one
has to put a specific number in a specific register of a computer in order to carry
out an operation with the assembly or machine instructions. It was the development
of high level programming languages, like Basic and Fortran, that made the wide-
spread use of sequential computers possible. Although codes written in lower level
assembly or machine languages are in general more efficient than those written in high
level programming languages, the benefits resulting from the ease of code development
and maintenance with high level languages significantly outweigh the loss in execution
speed. That is why nowadays we hardly see anyone developing large application codes
with low level assembly or machine languages.

An evolution process similar to that from machine language to high level programming language is repeating in parallel processing, but this time from manual
parallelization to automatic parallelization. The following is a list of the tools for
automatic or semi-automatic parallelization of sequential codes:

- KAP/KSR-1 [108], developed by Kuck and Associates in USA and implemented
  on Kendall Square KSR-1 parallel computers, supports Fortran and parallelizes
  sequential codes written in Fortran.

- KAP/Cray [147], developed by Kuck and Associates and implemented on Cray
  supercomputers, supports Fortran. In addition to parallelizing loops in sequential codes, it also makes necessary transformation of the sequential code to take
  advantage of the Cray architecture.

- FPP/Cray [42], developed by Cray research, supports Fortran and parallelizes
  sequential code by executing do-loops in parallel.

- Forge 90 [10], developed by Applied Parallel Research, Inc. in USA, supports
  Fortran 77, Fortran 90 and HPF (High Performance Fortran) and runs on Cray
  Y-MP, iPSC/860, Paragon, CM2, CM5, and clustered workstations. This is an
  interactive tool for parallelizing do-loops in a sequential program, so that users
  have easy access to the tool and better control over how a sequential code is
  parallelized.

- Parascope [25, 26], developed at Rice University in USA, supports Fortran 77
  and runs on clusters of workstations (Sun, IBM RS 6000). This is basically an

analysis tool that can be used by users to analyze data dependence in loops and optimize codes using different strategies, for example, loop reordering, unrolling, and strip mining.

- PAT (Parallelization AssistanT) [8, 9, 162], developed at Georgia Institute of Technology in USA, supports Fortran 77 and generates parallel codes for Cray Y-MP, IBM 3090, and KSR-1. This is also an interactive tool, so that users have control over the parallelization decision and can follow the progress of the parallelization process. The tool can be used to analyze data dependence and parallelize loops in sequential Fortran codes.

The six tools listed above represent only a small part of the research currently being carried out for developing automatic parallel compilers. The first three tools are non-interactive, while the other three are interactive. We have discussed, in Section 4.4, the numerical results obtained using the automatic parallelization tool KAP on KSR-1 parallel computers. Although the speedups achieved with automatic parallelization is not nearly as good as that with manual parallelization, it is certain that the wide-spread use of parallel computers in practical applications relies on the maturity of automatic parallelization tools. At the present, it is not unusual for researchers to spend six months to a year to port a large application code (with tens of thousands of lines) to a new parallel computer. For example, the author of this book participated in porting a computational fluid dynamics code and a reservoir simulation code to two different architectures. The first code has only about six thousand lines of Fortran and was originally developed for solving incompressible Navier-Stokes equations. It took about three months to parallelize the code and port it to the Intel iPSC/860 to generate correct results, and another three months to fine tune the code for improved efficiency. The other code is for reservoir simulations and has 32,000 lines of Fortran. This code was ported to KSR-1 parallel computers. Since KSR-1 has a shared-memory-like environment, it took only several days to get the code running and producing correct results. However, the period for fine tunning and improved performance lasted for about six months. There are usually two steps in the porting process. The first step is to reproduce the correct results as that generated by sequential computers. As discussed in previous sections, parallel implementation of a code involves domain decomposition, data sharing, and message exchange. Therefore, in many circumstances, it is not a trivial task to make the parallel code produce the correct results as those obtained on sequential computers. The second step is the fine-tuning to improve the performance and to fully utilize the features of the given parallel architecture. Considering the amount of work involved in porting large codes to parallel computers, we can see that an automatic parallelization tool that can put a sequential code runing right away on a parallel computer can save tremendous amount of work and time, even if the automatically parallelized code might run several times slower than the manually fine-tuned code which, as we have discussed before, may take months to implement. This is particularly true for the cases when the computational results are needed in a timely manner.

The main problems existing in the current available automatic parallelization tools are the following:

1. Lack of the capability for complicated data dependence analysis. The decision to parallelize a loop depends on the results of data or control dependence analysis. If there is no data dependence in a do-loop, then all the loops can be executed simultaneously on multiple processors. That is how most of the current automatic parallelization tools parallelize sequential codes. Therefore, accurate and efficient data dependence analysis is the key for automatically parallelizing complicated sequential codes. For example, the KAP on KSR-1 parallel computers can be used to automatically parallelize simple loops, but for complicated do-loops involving either arrays that need to be kept private for different processors or subroutine calls inside the loop body, it will not be able to make the data arrangement or to detect the data dependence further into subroutines called within the loop body. As a result, those loops which probably carry most of the computations will not be parallelized.

2. Lack of the capability for determine the granularity of parallel tasks to optimize the performance. This problem is also related to the first problem of data dependence analysis. In scientific computations, particularly with the MIMD architecture and physically distributed memory, coarse-grained parallel processing is more efficient than fine-grained implementations. If the automatic parallelization tool is not able to detect parallelism at coarse-grain level, for example nested loops with subroutine calls, then the parallelization has to be at a very fine grain level represented by simple non-nested do-loops without any subroutine calls. On the other hand, even if the automatic parallelization tool is able to detect high level, coarse-grained parallelism, it still may not be able to determine the grain size for parallelization since the problem size is not known until the input data is given to the program at run time. Therefore, a good automatic parallelization tool must be able to analyze data dependence in a complicated program, detect parallelism at coarse grain level, and determine the size of parallelized tasks based on the run-time input data to optimize the performance.

The improvement and progress of research in this area will depend on the accomplishments not only in parallel computing itself, but also in the general computer science as well, including compiler technology, algorithm analysis, and artificial intelligence.

6.4   Distributed Computing Network

A distributed computing network is a collection of computer systems connected by
a network that can be used as a coherent and flexible computational resource. It is
very common nowadays to see many computer systems, from main frames to PCs,
connected together by local and global networks. A key factor in a distributed parallel
computing network is the software environment that enables users to use all resources
in the network concurrently for solving a given large scale problem.

6.4.1   Homogeneous Computing Network

All computer systems are exactly the same in a homogeneous distributed computing
network. For example, a laboratory or a research center may have hundreds of the
same Sun workstations connected by a local network. A large application problem can
be decomposed into subproblems which are then processed on different workstations
as shown in Fig. 6.4.

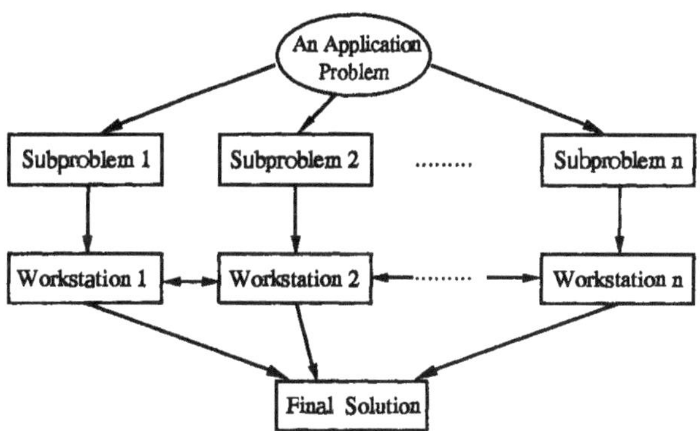

Figure 6.4: A large application problem can be decomposed into subproblems and solved concurrently on a cluster of workstations in a homogeneous computing network.

The information exchange between different workstations can be routed through
the local network. As mentioned before, the key part of this computing environment
is the software tools that handle the job scheduling and communications. This type
of parallel computing network not only provides us a new powerful computing tool,

but also utilizes the resources which have been wasted since most of those desktop workstations stay idle after working hours.

6.4.2 Heterogeneous Computing Network

Going one step further from the homogeneous computing network, we can consider different computer systems, from the desktop workstations to the fastest supercomputers, as a single computing resource for solving application problems. Fig. 6.5 depicts such a heterogeneous distributed computing network.

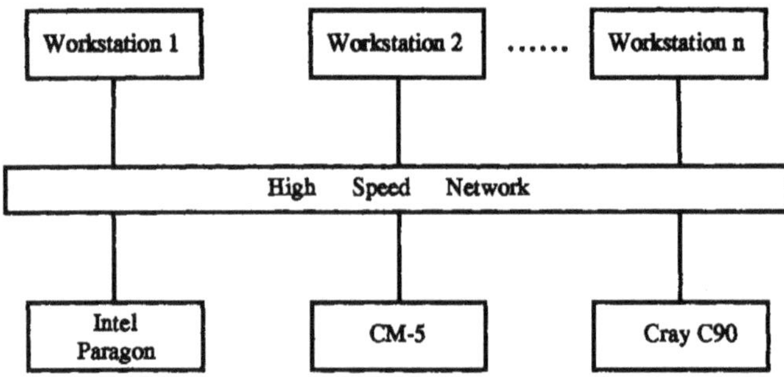

Figure 6.5: A heterogeneous computing network

In this kind of computing network, parallel processing must be interpreted in a much wider sense. As shown in Fig. 6.5, different computer systems are connected as a single concurrent computing resource. Among them are high resolution graphics workstations and parallel computers capable of performing parallel jobs with different granularities. Technically, a large application problem, or the so-called "Grand Challenge" problems in the areas like turbulence modeling, global weather study, or new drug design, can be decomposed into parts that can be best handled by different computers concurrently. For example, the visualization part can be handled by the graphics workstations, while the computationally intensive parts can be assigned to the best suitable parallel computers. Each part assigned to a particular parallel computer can also be processed by the multiprocessors on the parallel computer, thus introducing more than one level of parallelism, as shown in Fig. 6.6.

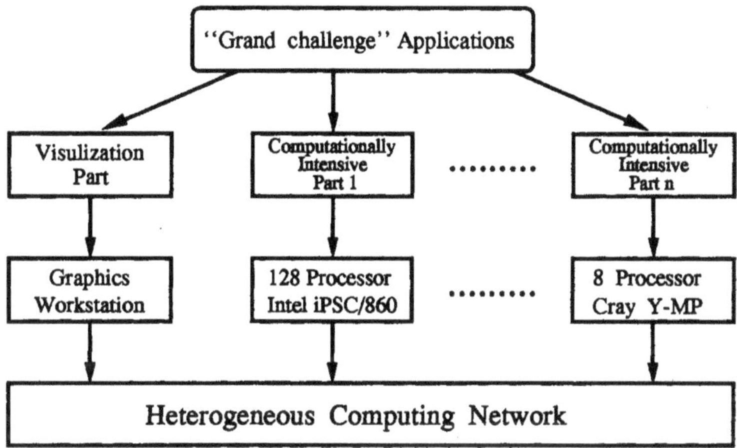

Figure 6.6: An application problem being processed on a heterogeneous computing network. Different parts of the problem are handled by different computers.

6.4.3　Software Tools

There are now several environments available which allow users to use networked computers as a unified concurrent computing resource. Here we briefly describe the PVM (Parallel Virtual Machine) system developed at the Oak Ridge National Laboratory in the United States.

The PVM is a software system that enables a collection of heterogeneous computer systems to be used as a coherent and flexible concurrent computational resource. The individual machine may be a shared- or distributed-memory parallel computer, a vector supercomputor, or a specialized graphics or scalar workstation that is interconnected by a variety of networks. User programs, written in C or FORTRAN, are provided access to the PVM in the form of library routines for functions such as process initiation, message transmission and reception, and synchronization. A programmer may optionally control the execution location of specific parts of the application problem being solved on the computing network. The PVM system transparently handles message routine, data conversion for incompatible computer architectures, and other tasks necessary for operations in a heterogeneous computing network. It is ideally suited for concurrent applications composed of many interrelated snb-algorithms that exploit specific strengths of individual machines on a network. It has been used for molecular dynamics simulations, superconductivity studies, distributed fractal computations and matrix algorithms. Interested users can get the PVM system free of charge by sending electronic mail to netlib@ornl.gov with the message "send index from PVM". Users guide and installation notes are included in the package. Currently supported architectures in the system include

- Sun workstations

- Sparc workstations

- IBM/RS6000

- Sequent Symmetry

- Cray

- Intel iPSC/860

- Intel iPSC/2

- CM2

- Alliant FX18

- Stardent Titan

and the list is growing constantly.

With the PVM system, an application program is decomposed into components that are subtasks carrying moderately large chunks of computational work. During execution, multiple instances of each component may be initiated and processed by different machines on the computing network, as shown in Fig. 6.7.

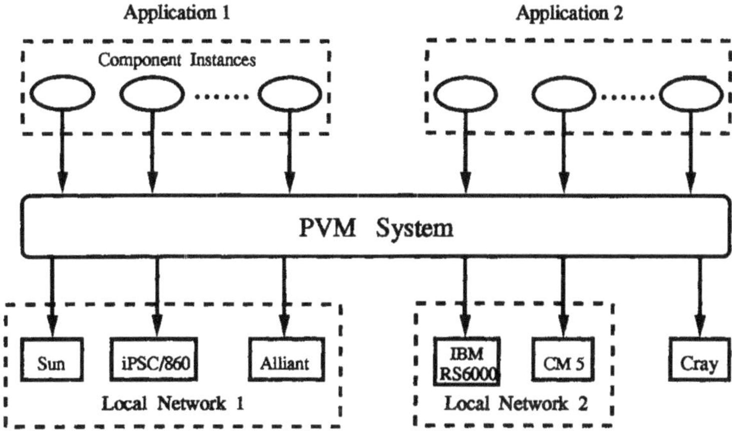

Figure 6.7: Execution model in the PVM environment

Different architectures on the network can be accessed in three different modes: the transparent mode in which component instances of an application program are automatically assigned to the most appropriate sites by the PVM system; the architecture-dependent mode in which the user may indicate specific architectures (for example Sun workstations) on which the particular components are to be executed; and the low-level mode in which a particular machine (for example Sun workstation number 2) may be specified for a specific component of the program. The initiation and termination of the component processes across the network, as well as communication and synchronization between processes can all be accomplished by calling the standard interface routines in the PVM library.

In addition to the PVM system, there are many other software tools developed in the last several years for heterogeneous network computing. The following is an incomplete list of software environments that will allow users to use distributed computing resources, including workstations, vector supercomputers, and massively parallel computers, as a coherent computer to solve a large scale problem:

- Linda [4, 63], developed by the Scientific Computing Associates Inc. in USA, supports Fortran and C programming language and runs on iPSC/2, SUN, IRIS, IBM RS-6000, Apollo, Encore and Sequent.

- Express [145], developed by ParaSoft in USA, supports C and Fortran programming language and runs on Crays, Intel hypercubes, IBM 3090, nCUBE, Transputers, and network of HP 9000/700, IBM RS-6000, SUN, and SGI workstations.

- P4 [24], developed at Argonne National Laboratory in USA, supports C and Fortran programming languages and runs on CM-5, Intel hypercubes, IBM 3090, Cray, Alliant Fx/8, Sequent, nCUBE, and network of IBM RS-6000, SUN, and SGI workstations.

- TCGMSG (Theoretical Chemistry Group Message Passing Toolkit) [77], developed at the Battelle Pacific Northwest Laboratory in USA, runs on Intel hypercubes, KSR-1, Alliant, network of SUN, DEC, HP, SGI and IBM RS-6000 workstations.

- CPS (Cooperative Processes Software) [51, 133], developed at the Fermi National Accelerator Laboratory in USA, runs on network of SGI, IBM RS-6000, DEC, SUN, HP, and MIPS workstations.

- PARMACS [17, 82, 83], developed in Germany, runs on Intel hypercubes, nCUBE Parsytec GC, Cray Y-MP, and network of DEC, IBM RS-6000, SGI, and SUN workstations.

- APPL (Application Portable Parallel Library) [150], developed at the NASA Lewis Research Center in USA, runs on Intel hypercubes, Alliant Fx/8, and network of SGI, SUN, IBM RS-6000 workstations.

- PARTI (Parallel Automated Runtime Toolkit at ICASE) [35, 43], developed at the Institute of Computer Applications in Science and Engineering (ICASE), runs on Intel hypercubes, nCUBE, CM-5, and network of workstations. This tool is built on top of the PVM and Express.

- SPPL (The Stuttgart Parallel Processing Library) [206], developed at University of Stuttgart in Germany, runs on a network of SUN, DEC, and IBM RS-6000 workstations. This tool only supports C programming language.

As can be seen from the above list, some of these software environments also run on vector supercomputers with shared memory and massively parallel computers with distributed memory, in addition to network of workstations. Therefore, these tools not only enable users to use distributed computing resources as a coherent computer, but also greatly increase the portability of parallel programs developed in these environments. For example, the tool TCGMSG provides users a message passing library that can be used on both shared and distributed memory parallel computers. A user can develop and implement codes on different architectures supported by the TCGMSG using the same message passing subroutine. The machine dependent part has been taken care of by the implementations of the TCGMSG on different architectures.

Similar to the development in programming languages discussed in Section 6.2, there is also a need for better organized and more widely supported effort to develop an efficient software environment for distributed computing network. With the participation of computer vendors, software developers and users, the next generation of distributed computing environment could be a powerful tool for developing portable parallel programs and using various computing resources more efficiently.

6.5 Metacenters and Metacomputers

Based on the concept of distributed computing network discussed in Section 6.4, the four national supercomputing centers supported by the US National Science Foundation have proposed in 1992 the formation of a National Metacenter for high performance computing. The creation of such a center extends distributed computing from in-house local network to national or even international network. The goal of the Metacenter is to give scientists and engineers the capability to move their computational problems directly to appropriate computers without regard to where the computers are located or where the researchers have allocations of supercomputer time; to develop a national file system to give researchers direct access to their files regardless of where these files are located; and to design a common user interface that allows researchers to use the same command to accomplish tasks on all different computing platforms at all centers. Each of the four National Supercomputer Centers shown geographically in Fig. 6.8 will contribute to the Metacenter its unique strength and specialty in architecture, software, and scientific applications.

- Cornell Theory Center. Located at Cornell University, the center hosts the latest IBM scalable parallel computer SP-1 and the Kendall Square shared virtual memory parallel computer KSR-1. Inter-disciplinary research projects

Figure 6.8: The geographical locations of the four National Supercomputer Centers in the United States. 1. Cornell Theory Center (New York). 2. Pittsburgh Supercomputing Center (Pennsylvania). 3.₁ National Center for Supercomputer Applications (Illinois). 4. San Diego Supercomputer Center (California). The four centers which are about 5000 kilometers apart are forming a national metacenter to provide users a unified coherent computing resource.

conducted on the centers computing platforms cover physics, mathematics, and many areas of engineering.

- National Center for Supercomputing Applications. This center is located at University of Illinois at Urbana-Champaign. It has a 512-node CM-5 parallel computer from the Thinking Machine, and a large memory Convex C3880 which will serve as a national visualization server. The center's software development group is building a Metacenter standard for data exchange (HDF) and constructing a data communication library (DTM). Research projects currently being conducted on the center's computing platforms include areas of mathematics and computer science, biological and medical sciences, earth and environmental sciences, and astronomy.

- Pittsburgh Supercomputing Center. This is a collaboration of Carnegie Mellon University, University of Pittsburgh and Westinghouse Electric Corp. The center has pioneered heterogeneous computing as a paradigm for research in a supercomputing environment. In 1990, the center's Cray Y-MP and CM2 were connected using the first high-speed link, so that application programs can be distributed between vector/parallel MIMD and massively SIMD architectures. Significant speedup has been achieved using this strategy because different parts of a large application code may be executed more efficiently on different architectures. Currently, the Center is developing the first single-vendor, tightly-coupled heterogeneous supercomputing environment by connecting the latest 16-processor Cray C90 and Cray Research's scalable massively parallel machine T3D. The current research projects supported by the Center cover the areas of structural biology, protein and DNA sequence analysis, high performance imaging in biological research, and large scale environmental modeling.

- San Diego Supercomputer Center. This Center is located in the west coast on the campus of University of California at San Diego. It currently hosts a massively parallel Inter Paragon supercomputer. Sustained Teraflops (10^{12} floating point operations per second) performance is expected by 1995. The center's software team is developing the computational and communications infrastructure for the future Metacenter, which includes high speed gigabit-per-second network testbeds, network performance analysis, network monitoring and measurement tools, and scientific databases.

It is clear that the Metacenter formed by the four National Supercomputer Centers provides researchers easy access of a diversity of resources, which may not be available at any single university, research laboratory, or supercomputer center. The resources provided by the Metacenter make it possible for researchers to develop complete field simulation systems for solving complicated engineering problems, and to improve major product (like airplane, missile, and automobile) designs in a timely manner. In such systems, efficient solution algorithms for partial differential equations is the core part that generates numerous amount of solution data. This part is usually the most computationally intensive and requires massively parallel computers. Other critical parts in a complete simulation system include grid generation for complex geometries and visualization of the solution data for accurate and fast analysis. The former is usually called preprocessing and the latter postprocessing. These two pars usually require enhanced graphics processing capability, high speed data transmission, mass data storage, and efficient data management. An integrated simulation system including all these parts, for example the computational field simulation system being developed at the NSF supported Engineering Research Center of Mississippi State University, depends heavily on the development of hardware and software resources that can be offered only by Metacenters.

It can be expected that the formation and expansion of Metacenters, and the related research in software and hardware, will lead the development of high performance computing into the next century. Coupled with this development, the complete simulation systems with efficient PDE solution algorithms will become truly indispensable tools for solving complex real world application problems.

Bibliography

[1] J.C. Adams, W.S. Brainerd, J.T. Martin, B.T. Smith, and J.L. Wagener. *FOR-TRAN 90 Handbook—Complete ANSI/ISO Reference*. McGraw-Hill, 1992.

[2] L. Adams and H. Jordan. Is SOR color-blind? *SIAM J. Sci. Stat. Comput.*, 7:490–506, 1985.

[3] L. Adams, R. LeVeque, and D. Young. Analysis of the SOR iteration for the 9-point Laplacian. *SIAM J. Numer. Anal.*, 1987.

[4] S. Ahuja, N. Carriero, and D. Gelernter. Linda and friends. *IEEE Computer*, 26–34, 1986.

[5] M. Alef. Concepts for efficient multigrid implementation on SUPRENUM-like architectures. *Parallel Computing*, 17:1–17, 1991.

[6] G. Amdahl. Validity of the single-processor approach to achieving large scale computing capabilities. In *Proc. AFIPS Conf.*, pages 483–485, 1967.

[7] W.F. Ames. *Numerical Methods for Partial Differential Equations*. Academic Press, San Diego, 1992.

[8] B. Appelbe and K. Smith. Start/Pat: A parallel-programming toolkit. *IEEE Software*, 29-38, 1989.

[9] B. Appelbe, K. Smith, and K. Stirewalt. PATCH—A new algorithm for rapid incremental dependence analysis. *ICS*, 424–432, 1991.

[10] Applied Parallel Research. *FORGE 90, Version 8.0, Baseline System User's Guide*, 1992.

[11] C. Ashcraft. Domain decoupled incomplete factorizations. Applied mathematics technical report, ETA-TR-49, Boeing Computer Services, 1987.

[12] O. Axelsson. A class of iterative methods for finite element equations. *Comput. Methods Appl. Mech. Eng.*, 9:123–137, 1976.

[13] K. Aziz and A. Settari. *Petroleum Reservoir Simulation.* Applied Science Publishers Ltd., London, 1979.

[14] D.H. Bailey, E. Barszcz, L. Dagum, and H.D. Simon. NAS parallel benchmark results. RNR technical report, RNR-92-002, NASA Ames Research Center, Moffett, CA 94035, 1992.

[15] P. Bastian, J. Burmeister, and G. Horton. Implementation of a parallel multigrid method for parabolic partial differential equations. In W. Hackbusch, editor, *Proceedings of the 6th GAMM Seminar*, pages 18–27, Vieweg. Braunschweig, 1990.

[16] T. Bemmerl and A. Bode. An integrated tool environment for programming distributed memory multiprocessors. In A. Bode, editor, *Distributed memory computing, Lecture Notes in Computer Science*, volume 487, pages 130–142. Springer-Verlag, 1991.

[17] L. Bomans, R. Hempel, and D. Roose. The Argonne/GMD macros in Fortran for portable parallel programming and their implementation on the Intel iPSC/2. *Parallel Computing*, 15:119–132, 1990.

[18] T. Brandes. Efficient data parallel program without explicit message passing for distributed memory multiprocessors. GMD Techinical Report TR92-4, 1992.

[19] A. Brandt. Multigrid solvers on parallel computers. In M. Schultz, editor, *Elliptic Problem Solvers.* Academic Press, New York, 1981.

[20] A. Brandt. Multigrid techniques: 1984 guide with applications to fluid dynamics. *GMD-Studie Nr.*, 85, 1984.

[21] P. Brezany, B. Chapman, and H. Zima. Automatic parallelization for GENESIS. Technical Report ACPC/TR 92-16, Austrian Center for Parallel Computation, 1992.

[22] W.L. Briggs. *A Multigrid Tutorial.* SIAM, Philadelphia, 1987.

[23] J. Burmeister. Paralleles Lösen diskreter parabolischer Probleme mit Mehrgittertechniken. Master's thesis, Universität Kiel, 1985.

[24] R. Butler and E. Lusk. User's guide to the P4 parallel programming system. Technical Report ANL-92/17, Argonne National Laboratory, Mathematics and Computer Science Division, 1992.

[25] D. Callahan, K.D. Cooper, K. Kennedy, and L. Torczon. Interprocedural constant propagation. *ACM SIGPLAN Notices*, 21(7):152–161, 1986.

[26] A. Carle, K.D. Cooper, R.T. Hood, K. Kennedy, L. Torczon, and S.K. Warren. A practical environment for scientific programming. *IEEE Computer*, 1987.

[27] T. Chan. On the implementation of Kernel Numerical Algorithms for computational fluid dynamics on hypercubes. In M. Heath, editor, *Hypercube Multiprocessors*, pages 747–755. SIAM, Philadelphia, 1987.

[28] T. Chan. Domain decomposition algorithms and computational fluid dynamics. *Int. J. Supercomputer Applications*, 2:72–83, 1988.

[29] T. Chan and D. Resasco. Hypercube implementation of domain decomposed fast Poisson solvers. In Heath [27], pages 738–746.

[30] T. Chan and Y. Saad. Multigrid algorithms on the hypercube multiprocessor. *IEEE Trans. Comput.*, 35:969–977, 1986.

[31] T. Chan and R. Schreiber. Parallel networks for multigrid algorithms: Architecture and complexity. *SIAM J. Sci. Comput.*, 6:698–711, 1985.

[32] T. Chan and R. Tuminaro. Design and implementation of parallel multigrid algorithms. In S. McCormick, editor, *Multigrid Methods: Theory, Applications, and Supercomputing*, pages 101–115. Marcel Dekker Inc., New York, 1988.

[33] B. Chapman, P. Mehrota, and H. Zima. Programming in Vienna Fortran. *Scientific Programming*, 1(1), 1992.

[34] B. Chapman, P. Mehrotra, and H. Zima. Vienna FORTRAN — A Fortran language extension for distributed memory multiprocessors. In J. Saltz and P. Mehrotra, editors, *Compilers and Runtime Software for Scalable Multiprocessors*. Elsevier Science Publishers B.V., Amsterdam, 1991.

[35] C. Chase, K. Crowley, J. Saltz, and A. Reeves. Parallelization of irregularly coupled regular meshes. In *Proceedings of the Sixth International Conference on Supercomputing*. Washington DC, 1992.

[36] D.Y. Chen. A survey of parallel programming languages and tools. Report RND-93-005, NASA Ames Research Center, Moffett Field, California, 1993.

[37] A.T. Chronopoulos and C.W. Gear. On the efficient implementation of preconditioned s-step conjugate gradient methods on multiprocessors with memory hierarchy. *Parallel Computing*, 11:37–53, 1989.

[38] A.T. Chronopoulos and C.W. Gear. S-step iterative methods for symmetric linear systems. *J. Comput. Appl. Math.*, 25:153–168, 1989.

[39] P. Concus, G. Golub, and D. O'Leary. A generalized conjugate gradient method for the numerical solution of elliptic partial differential equations. In J. Bunch and D. Rose, editors, *Sparse Matrix Computations*, pages 309–322. Academic Press, New York, 1976.

[40] P. Concus and D. O'Leary. Block preconditioning for the conjugate gradient method. *SIAM J. Sci. Stat. Comput.*, 6:220–252, 1985.

[41] R. Courant and D. Hilbert. *Methods of Mathematical Physics*, volume 2. Interscience, New York, 1962.

[42] Cray Research, Inc. *CF77 Compiling System, Volume 4: Parallel Processing Guide.*

[43] R. Das, D.J. Mavriplis, J. Saltz, S. Gupta, and R. Ponnusamy. The design and implementation of a parallel unstructured Euler solver using software primitives (AIAA-92-0562). In *Proceedings of the 30th Aerospace Sciences Meeting.* Reno NV, 1992.

[44] Y. Deng, J. Glimm, and D.H. Sharp. Perspectives on parallel computing. *Daedalus*, 12:31–52, 1992.

[45] J.J. Dongarra, I.S. Duff, D.C. Sorensen, and H.A. Van der Vorst. *Solving Linear Systems on Vector and Shared Memory Computers.* SIAM, Philadelphia, 1991.

[46] T.H. Dunigan. Kendall Square multiprocessor: Early experiences and performance. Technical Report ORNL/TM-12065, Oak Ridge National Laboratory, 1992.

[47] H.C. Elman and E. Agrón. Ordering techniques for the preconditioned conjugate gradient method on parallel computers. *Computer Physics Communications*, 53:253–269, 1989.

[48] D.J. Evans and C. Li. Successive underrelaxation (SUR) and generalized conjugate gradient (GCG) methods for hyperbolic difference equations on a parallel computer. *Parallel Computing*, 16:207–219, 1990.

[49] R.E. Ewing. The mathematics of reservoir simulations. In *Frontiers in Applied Mathematics*, volume 1. SIAM, Philadelphia, 1983.

[50] R.E. Ewing. Large-scale computing in reservoir simulations. *Int. J. Supercomput. Appl.*, 2(4):44–53, 1988.

[51] R. Fausey, P. Wolbers, and U. Yeager. *CPS & CPS Batch Reference Guide*. Fermi Laboratory, Computing Division, GA0008.

[52] S. Filippone, M. Marrone, and G.R. Di Brozolo. Parallel preconditioned conjugate-gradient type algorithms for general sparsity structures. *Int. J. Computer Math.*, 40:159–167, 1992.

[53] R. Ford. Laminar to turbulent transition. Architecture and simulation implementation report, Center for Novel Computing, University of Manchester, Manchester, England, 1992.

[54] G. Fox, S. Hiranandani, K. Kennedy, C. Koelbel, U. Kremer, C.W. Tseng, and M.Y. Wu. Fortran D language specification. Dept. of Computer Science Technical Report TR90-141, Rice University, 1990.

[55] P.O. Frederickson and O.A. McBryan. Normalized convergence rates for the PSMG Method. *SIAM J. Sci. Statist. Comput.*

[56] P.O. Frederickson and O.A. McBryan. Parallel superconvergent multigrid. In S. McCormick, editor, *Multigrid Methods: Theory, Applications and Supercomputing*. Dekker, New York, 1988.

[57] P.O. Frederickson and O.A. McBryan. Recent developments for parallel multigrid. In U. Trottenberg and W. Hackbusch, editors, *Proceedings of the Third European Conference on Multigrid Methods*, 1990.

[58] D. Gannon and J. van Rosendale. On the structure of parallelism in a highly concurrent PDE solver. *J. Par. Dist. Comp.*, 3:106–135, 1986.

[59] P. Garabedian. *Partial Differential Equations*. John Wiley & Sons, New York, 1964.

[60] V. Gärtel. Parallel multigrid solver for 3D anisotropic elliptic problems. Technical Report 390, Gesellschaft für Mathematik und Datenverarbeitung, 1989.

[61] C. Gear. *Numerical Initial Value Problems in Ordinary Differential Equations.* Prentice-Hall, Englewood Cliffs, 1971.

[62] C. Gear and F. Juang. The speed of waveform methods for ODEs. In R. Spigler, editor, *Applied and Industrial Mathematics*, pages 37–48. Kluwer, Boston, 1991.

[63] D. Gelernter, N. Carriero, S. Chandran, and S. Chang. Parallel programming in Linda. In *Proceedings of the 1985 International Conference on Parallel Processing*, pages 255–263, 1985.

[64] A. George. Nested dissection of a regular finite element mesh. *SIAM J. Numer. Anal.*, 10:345–363, 1973.

[65] J. Glimm and O.A. McBryan. In R.E. Ewing, editor, *The Mathematics of Reservoir Simulations*, pages 107–160. SIAM, Philadelphia, 1983.

[66] G.H. Golub and C.F. van Loan. *Matrix computations.* The Johns Hopkins University Press, Baltimore, 2nd edition, 1989.

[67] A. Greenbaum. A multigrid method for multiprocessors. *Appl. Math. Comput.*, 19:75–88, 1986.

[68] W. Gropp and D. Keyes. Domain decomposition on parallel computers. In T. Chan et al., editors, *Domain Decomposition Methods*. SIAM, Philadelphia, 1989.

[69] W. Gropp and E. Smith. Computation fluid dynamics on parallel processors. *Computer & Fluids*, 18:289, 1990.

[70] C.E. Grosch. Poisson solvers on large array computer. In B.L. Buzbee and J.F. Morrison, editors, *Proceedings 1978 LANL Workshop on vector and parallel processors*, 1978.

[71] C.E. Grosch. Performance analysis of Poisson solvers on array computers. Report TR 79-3, Old Dominion University, Norfolk, VA, 1979.

[72] J. Gustafson. Fixed time, tiered memory, and superlinear speedup. In D.W. Walker and Q.F. Stout, editors, *Proceedings of the Fifth Distributed Memory Computing Conference*, pages 1255–1260. IEEE Computer Society Press, 1990.

[73] J. Gustafson, G. Montry, and R. Benner. Development of parallel methods for a 1024-processor hypercube. *SIAM J. Sci. Stat. Comput.*, 9(4):609–638, 1988.

[74] W. Hackbusch. Parabolic multi-grid methods. In R. Glowinski and J.R. Lions, editors, *Comp. Meth. in Appl. Sci. and Eng., VI.* North Holland, Amsterdam, 1984.

[75] W. Hackbusch. *Multi-grid Methods and Applications.* Springer Verlag, Berlin, 1987.

[76] W. Hackbusch and U. Trottenberg. Multigrid methods. In *Lecture Notes in Mathematics*, volume 960, Springer-Verlag. Berlin, 1982.

[77] R.J. Harrison. Moving beyond message passing: Experiments with a distributed-data model. Technical report, Argonne National Laboratory, 1991.

[78] M.T. Heath, G.A. Geist, and J.B. Drake. Early experience with the Intel iPSC/860 at Oak Ridge National Laboratory. *Int. J. of Supercomputer Applications*, 5(2):10–26, 1991.

[79] M.T. Heath and C.H. Romine. Parallel solution of triangular systems on distributed-memory message-passing multiprocessors. *SIAM J. Sci. Stat. Comput.*, 9:558–587, 1988.

[80] D. Heller. On the efficient computation of recurrence relations. Technical report, Institute for Computer Applications in Science and Engineering, NASA Langley Research Center, Hampton, VA, 1974.

[81] D.P. Helmbold and C.E. McDowell. Modeling speedup(n) greater than n. *Proceedings of the International Conference on Parallel Processing*, 3:219–225, 1989.

[82] R. Hempel. The ANL/GMD macros (PARMACS) in Fortran for portable parallel programming using the message passing programming model. *User's Guide and Reference Manual*, 1991. Version 5.1.

[83] R. Hempel, H.C. Hoppe, and A. Supalov. PARMACS 6.0 library interface specification. GMD Internal Report, 1992.

[84] R. Hempel and A. Schüller. Experiments with parallel multigrid using the SUPRENUM communications library. *GMD-Studie Nr.*, 141, 1988.

[85] R. Herbin, S. Gerbi, and V. Sonnad. Parallel implementation of a multigrid method on the experimental ICAP supercomputer. *Appl. Math. Comput.*, 27:281–312, 1988.

[86] M. Hestenes. The conjugate gradient method for solving linear systems. In *Proc. Sixth Symp. Appl. Math.*, pages 83–102, McGraw-Hill. New York, 1956.

[87] M. Hestenes and E. Stiefe. Methods of conjugate gradients for solving linear systems. *J. of Research of National Bureau of Standards*, B49:409–436, 1952.

[88] High Performance Fortran Forum. *High Performance Fortran Language Specification, Version 1.0*, 1993.

[89] S. Hiranandani, K. Kennedy, and C.W. Tseng. Compiling Fortran D for MIMD distributed-memory machines. *Communications of the ACM*, 35(8):66–80, 1992.

[90] C. Ho and S. Johnsson. Optimizing tridiagonal solvers for alternating direction methods on boolean cube multiprocessors. *SIAM J. Sci Stat. Comput.*, 11(3):563–592, 1990.

[91] R.W. Hockney and C.R. Jesshope. *Parallel Computers: Architecture, Programming, and Algorithms*. Adam Hilger Ltd., Bristol, UK, 1981.

[92] G. Horton. Time-parallel multigrid solution of the Navier-Stokes equations. In C.A. Brebbia, D. Howard, and A. Peters, editors, *Proc. Conf. on Appl. of Supercomputers in Eng. II*, pages 435–445, Comp. Mech. Publ. Boston, 1991.

[93] G. Horton and R. Knirsch. A space- and time-parallel multigrid method for time-dependent p.d.e.s. Technical Report 2, Institut für Mathematische Maschinen und Datenverarbeitung der Friedrich-Alexander-Universität Erlangen-Nürnberg, 1992.

[94] G. Horton and R. Knirsch. A time-parallel multigrid-extrapolation method for parabolic partial differential equations. *Parallel Computing*, 18:21–29, 1992.

[95] K. Hwang. *Advanced Computer Architecture with Parallel Programming*. McGraw-Hill, preliminary edition, 1993.

[96] Intel Supercomputer Systems Division. *iPSC/2 and iPSC/860 Concurrent Programming, n. 2*, 1991.

[97] Intel Supercomputer Systems Division. *Intel iPSC/860 user training workshop manual*, 1992.

[98] ISO/IDE. Information technology—Programming languages—Fortran. *International Standard*, 1991. Reference number ISO/IEC 1539: 1991(E).

[99] C.P. Jackson and P.C. Robinson. A numerical study of various algorithms related to the preconditioned conjugate gradient method. *Int. J. Num. Methods Eng.*, 21:1315–1338, 1985.

[100] S. Johnsson. Solving tridiagonal systems on ensemble architectures. *SIAM J. Sci. Stat. Comput.*, 8(3):354–392, 1987.

[101] S. Johnsson, Y. Sadd, and M. Schultz. Alternating direction methods on multiprocessors. *SIAM J. Sci. Stat. Comput.*, 8(5):686–700, 1987.

[102] H. Jordan. The force. ECE Tech. Report 87-1-1, 1987.

[103] H.F. Jordan, M.S. Benten, N.S. Arenstorf, and A.V. Ramanan. *Force User's Manual*, 1989.

[104] M. Kadiöglu and S. Mudrick. On the implementation of the GMRES(m) method to elliptic equations in meteorology. *J. Computational Physics*, 102:348–359, 1992.

[105] L. Kang. *Parallel Algorithms and Domain Decomposition*. Wuhan University Press, Wuhan, China, 1987.

[106] A.H. Karp and H.P. Flatt. Measuring parallel processor performance. *Parallel Comput.*, 17:1111–1130, 1991.

[107] Kendall Square Rearch. *KSR Technical Summary*. Waltham, MA, USA, 1991.

[108] Kendall Square Research. *KSR Parallel Programming*. Waltham, MA, USA, 1991.

[109] R. Knirsch. Implementierung eines parallelen Mehrgitterverfahrens zur Lösung instationärer partieller Differentialgleichungen. Master's thesis, Universität Erlangen-Nürnberg, 1990.

[110] P.M. Kogge. Parallel solution of recurrence problems. *IBM Journal of Research and Development*, 18:138–148, 1974.

[111] P.M. Kogge and H.S. Stone. A parallel algorithm for the efficient solution of a general class of recurrence equations. *IEEE Trans. Comput.*, C-22:786–793, 1973.

[112] A. Krechel, H. Plum, and K. Stüben. Parallel solution of tridiagonal linear system. In F. André and J. Verjus, editors, *Hypercube and distributed computers*, pages 49–64. North-Holland, Amsterdam, 1989.

[113] D.J. Kuck. *The Structure of Computers and Computations*, volume 1. John Wiley & Sons, New York, 1978.

[114] M. Kumar. Measuring parallelism in computation intensive scientific/engineering applications. *IEEE-TC*, 37:1088–1098, 1988.

[115] J. Kuo, B. Levy, and B. Muskus. A local relaxation method for solving elliptic PDEs on mesh connected arrays. *SIAM J. Sci. Stat. Comput.*, 8:550–573, 1987.

[116] J. Lambert. *Computational Methods in Ordinary Differential Equations*. John Wiley & Sons, Chichester, 1973.

[117] L. Lapidus and G.F. Pinder. *Numerical Solution of Partial Differential Equations in Science and Engineering*. John Wiley & Sons, New York, 1982.

[118] C.E. Leiserson. Fat-trees: Universal networks for hardware-efficient supercomputing. *IEEE Transactions on Computing*, C-34(10):892–901, 1985.

[119] E. Lelarasmee. *The Waveform Relaxation Method for the Time Domain Analysis of Large Scale Nonlinear Systems*. PhD thesis, University of California, Berkeley, 1982.

[120] E. Lelarasmee, A. Ruehli, and A. Sangiovanni-Vincentelli. The waveform relaxation method for time-domain analysis of large scale integrated circuits. *IEEE Trans. on CAD of IC and Sys.*, 1(3):131–145, 1982.

[121] G. Li and T.F. Coleman. A new method for solving triangular systems on distributed-memory message-passing multiprocessors. *SIAM J. Sci. Stat. Comput.*, 10:382–396, 1989.

[122] C. Lubich and A. Ostermann. Multigrid dynamic iteration for parabolic equations. *BIT*, 27:216–234, 1987.

[123] D.G. Luenberger. *Introduction to Linear and Nonlinear Programming*. Addison-Wesley, New York, 1973.

[124] L. Mansfield. Damped Jacobi preconditioning and coarse grid deflation for conjugate gradient iteration on parallel computers. *SIAM J. Sci. Stat. Comput.*, 12(6):1314–1323, 1991.

[125] O.A. McBryan. Sequential and parallel efficiency of multigrid fast solvers. Technical report, University of Colorado Computer Science Department, 1990.

[126] O.A. McBryan and E.F. Van de Velde. Hypercube algorithms and implementations. *SIAM J. Sci. Comput.*, 8:s227–s287, 1987.

[127] O.A. McBryan, P.O. Frederickson, J. Linden, A. Schüller, K. Solchenbach, K. Stüben, C. Thole, and U. Trottenberg. Multigrid methods on parallel computers—A survey of recent developments. *Impact of Computing in Science and Engineering*, 3:1–75, 1991.

[128] G. Meurant. Practical use of the conjugate gradient method on parallel supercomputers. *Computer Physics Communications*, 53:467–477, 1989.

[129] U. Miekkala and O. Nevanlinna. Convergence of dynamic iteration methods for initial value problems. *SIAM J. Sci. Stat. Comput.*, 8(4):459–482, 1987.

[130] U. Miekkala and O. Nevanlinna. Sets of convergence and stability regions. *BIT*, 27:554–584, 1987.

[131] Manuals on: MMK, TOPSYS, DETOP, PATOP VISTOP. Technical report, Technische Universität München, 1991.

[132] N.M. Nachtigal, L. Reichel, and L.N. Trefethen. A hybrid GMRES algorithm for non-symmetric linear systems. *SIAM J. Matrix Anal. Appl.*, 13:796–825, 1992.

[133] T. Nash. High performance parallel local memory computing at Fermi-lab. In *Proceedings of WHP92 on Heterogeneous Processing*, 1993.

[134] A. Navarra. An application of GMRES to indefinite linear problems in meteorology. *Computer Physics Communications*, 53:321–327, 1989.

[135] O. Nevanlinna. Remarks on Picard-Lindelöf iteration, PART I. *BIT*, 29:328–346, 1989.

[136] O. Nevanlinna. Remarks on Picard-Lindelöf iteration, PART II. *BIT*, 29:535–562, 1989.

[137] O. Nevanlinna. Linear acceleration of Picard-Lindelöf iteration. *Numer. Math.*, 57:147–156, 1990.

[138] O. Nevanlinna. Power bounded prolongations and Picard-Lindelöf iteration. *Numer. Math.*, 58:479–501, 1990.

[139] A. Newton and A. Sangiovanni-Vincentelli. Relaxation-based electrical simulation. *SIAM J. Sci. Stat. Comput.*, 4(3):485–524, 1983.

[140] D.P. O'Leary. Parallel implementation of the block conjugate gradient algorithm. *Parallel Computing*, 5:127–139, 1987.

[141] T. Oppe and D. Kincaid. The performance of ITPACK on vector computers for solving large sparse linear systems arising in sample oil reservoir simulation problems. *Commun. Appl. Numer. Math.*, 3:23–30, 1987.

[142] J.M. Ortega. *Numerical analysis, a second course.* Academic Press, New York, 1972.

[143] J.M. Ortega. *Introduction to Parallel and Vector Solution of Linear Systems.* Plenum Press, New York, 1988.

[144] J.M. Ortega and R.G. Voigt. Solution of partial differential equations on vector and parallel computers. *SIAM Review*, 27:149–240, 1985.

[145] Parasoft Co. *Express C User's Guide*, 1990. Version 3.0.

[146] D. Parkinson. Parallel efficiency can be greater than unity. *Parallel Computing*, 3:261–262, 1986.

[147] D.M. Pase and K.E. Fletcher. Automatic parallelization: A comparison of CRAY FPP and KAI KAP/CRAY. NASA Ames NAS Technical Report, RND-90-010, 1990.

[148] D.W. Peaceman. *Fundamentals of numerical reservoir simulation.* Elsevier Scientific Publishing B.V., Amsterdam, 1977.

[149] D.W. Peaceman and H.H. Rachford. The numerical solution of parabolic and elliptic differential equations. *SIAM J.*, 3:28–41, 1955.

[150] A. Quealy, G.L. Gole, and R.A. Blech. Portable programming on parallel/networked computers using the application portable parallel library (APPL). *NASA TM*, 1993.

[151] L.F. Richardson. *Transaction of Royal Society*, A210, 1910.

[152] E. Rosti, E. Smirni, T.D. Wagner, A.W. Apon, and L.W. Dowdy. The KSR-1: Experimentation and modeling of poststore. Technical report, Department of Computer Science, Vanderbilt University, 1993.

[153] C.Z. Runge. *Mathematical Physics*, 56, 1908.

[154] Y. Saad. Krylov subspace methods on supercomputers. *SIAM J. Sci. Stat. Comput.*, 10:1200–1232, 1989.

[155] Y. Saad and M. Schultz. GMRES: A generalized minimal residual algorithm for solving nonsymmetric linear systems. *SIAM J. Sci. Statist. Comput.*, 7:856–869, 1986.

[156] F. Saied, C.T. Ho, L. Johnsson, and M. Schultz. Solving Schrodinger's Equation on the Intel iPSC by the alternating direction method. In M. Heath, editor, *Hypercube Multiprocessors*, pages 680–691. SIAM, Philadelphia, 1987.

[157] J.H. Saltz and V.K. Naik. Towards developing robust algorithms for solving partial differential equations on MIMD machines. *Parallel Computing*, 6(1), 1987.

[158] J.H. Saltz, V.K. Naik, and D.M. Nicol. Reduction of the effects of the communication delays in scientific algorithms on message passing MIMD architectures. *SIAM J. Sci. Stat. Comput.*, 8:S118–S134, 1987.

[159] A.H. Sameh, S. Chen, and D. Kuck. Parallel Poisson and biharmonic solvers. *Computing*, 17:219–230, 1976.

[160] L.R. Scott. Pfortran: A parallel dialect of Fortran. *Fortran Forum*, 11(3):20–31, 1992.

[161] F. Shakib, T.J. Hughes, and Z. Johan. A multi-element group preconditioned GMRES algorithm for non-symmetric systems arising in finite element analysis. *Computer Methods in Applied Mechanics and Engineering*, 75:415–456, 1989.

[162] K. Smith, B. Appelbe, and K. Stirewalt. Incremental dependence analysis for interactive parallelization. *ICS*, 330–341, 1990.

[163] N.A. Sobh. Preconditioned conjugate gradient and finite element methods for massively data-parallel architectures. *Computer Physics Communications*, 65:253–267, 1991.

[164] J.C. Strikwerda. *Finite Difference Schemes and Partial Differential Equations.* Wadsworth & Brooks/Cole Advanced Books & Software, Pacific Grove, California, 1989.

[165] R. Sumner. Transonic turbulent impinging jets code. Technical report, Center for Novel Computing, University of Manchester, Manchester, England, 1992.

[166] X.H. Sun and J.L. Gustafson. Toward a better parallel performance metric. *Parallel Computing*, 17:1093–1109, 1991.

[167] X.H. Sun and L. Ni. Another view on parallel speedup. In *Proc. Supercomputing'90*, pages 324–333. New York, 1990.

[168] X.H. Sun and L. Ni. Scalable problems and memory-bounded speedup. *J. of Parallel and Distributed Computing*, 19:27–37, 1993.

[169] L.H. Tan and K.J. Bath. Studies of finite element procedures—the conjugate gradient and GMRES methods in ADINA and ADINA-F. *Computers and Structures*, 40:441–449, 1991.

[170] C.A. Thole. Experiments with multigrid methods on the CalTech-hypercube. *GMD-Studie Nr.*, 103, 1985.

[171] C.A. Thole and U. Trottenberg. A short note on standard parallel multigrid algorithms for 3D problems. *Appl. Math. Comput.*, 27:101–115, 1988.

[172] J.F. Thompson, Z.U.A. Warsi, and C.W. Mastin. *Numerical Grid Generation — Foundations and Applications*. North-Holland, New York, 1985.

[173] R.S. Tuminaro. A highly parallel multigrid-like method for the solution of the Euler equations. *SIAM J. Sci. Stat. Comput.*, 13:88–100, 1992.

[174] H.A. Van Der Vorst. High performance preconditioning. *SIAM J. Sci. Stat. Comput.*, 10(6):1174–1185, 1989.

[175] S. Vandewalle. Activiteitenverslag 1985–1986. Interim Report N.F.W.O. (in Dutch), 1986.

[176] S. Vandewalle. Waveform relaxation methods for solving parabolic partial differential equations. In D. Walker and Q. Stout, editors, *Proceedings of Fifth Distributed Memory Computing Conference*, pages 575–584. IEEE Computer Society Press, 1990.

[177] S. Vandewalle. *The parallel solution of parabolic partial differential equations by multigrid waveform relaxation methods*. PhD thesis, Katholieke Universiteit Leuven, 1992.

[178] S. Vandewalle, R. Van Driessche, and R. Piessens. The parallel performance of standard parabolic marching schemes. *Int. J. High Speed Computing*, 3:1–29, 1991.

[179] S. Vandewalle and R. Piessens. The parallel waveform relaxation multigrid method. In G. Rodrigue, editor, *Proceedings of the Third SIAM Conference on Parallel Processing for Scientific Computing*, pages 152–156, SIAM. Philadelphia, 1989.

[180] S. Vandewalle and R. Piessens. Numerical experiments with nonlinear multigrid waveform relaxation on a parallel processor. *Applied Numerical Mathematics*, 8(2):149–161, 1991.

[181] S. Vandewalle and R. Piessens. Efficient parallel algorithms for solving initial-boundary value and time-periodic parabolic partial differential equations. *SIAM J. Sci. Stat. Comput.*, 13:1330–1346, 1992.

[182] S. Vandewalle and R. Piessens. On dynamic iteration methods for solving time-periodic differential equations. *SIAM J. Numer. Anal.*, 30:286–303, 1993.

[183] R. Varga. *Matrix Iterative Analysis*. Prentice Hall, Englewood Cliffs, New Jersey, 1962.

[184] H.H. Wang. A parallel method for tridiagonal equations. *ACM Trans. Math. Software*, 7:170–183, 1981.

[185] A. Wathen and D. Silvester. Fast iterative solution of stabilized Stokes systems, Part I: Using simple diagonal preconditioners. Technical Report NA-91-04, Computer Science Department, Stanford University, 1991.

[186] J.A. Wheeler and R.A. Smith. Reservoir simulation on a hypercube. *SPE Reservoir Engrg*, 544-548, 1990.

[187] J. White and A. Sangiovanni-Vincentelli. *Relaxation Techniques for the Simulation of VLSI Circuits*. Kluwer Academic Publishers, Boston, 1987.

[188] J. White, A. Sangiovanni-Vincentelli, F. Odeh, and A. Ruehli. Waveform relaxation: Theory and practice. *Trans. of the Soc. for Comp. Simulation*, 2(1):95–133, 1985.

[189] D.E. Womble. A time stepping algorithm for parallel computers. *SIAM J. Sci. Stat. Comput.*, 11(5):824–837, 1990.

[190] D.E. Womble and B.C. Young. Multigrid on massively parallel computers. In Walker and Stout [176], pages 575–584.

[191] P.T. Worley. The effect of time constraints on scaled speedup. *SIAM J. Sci. Stat. Comput.*, 11:838–858, 1990.

[192] P.T. Worley. Parallelizing across time when solving time-dependent partial differential equations. Technical Report ORNL/TM-11963, Oak Ridge National Laboratory, 1991.

[193] P.T. Worley. Parallelizing across time when solving time-dependent partial differential equations. In D. Sorensen, editor, *Proceedings of the Fifth SIAM Conference on Parallel Processing for Scientific Computing*. SIAM, 1991.

[194] D.P. Young. *Iterative Solution of Large Linear Systems*. Academic Press, New York, 1971.

[195] D.P. Young, R.G. Melvin, F.T. Johnson, J.E. Bussoletti, L.B. Wigton, and S.S. Samant. Application of sparse matrix solvers as effective preconditioners. *SIAM J. Sci. Stat. Comput.*, 10:1186–1199, 1989.

[196] Q. Zhang. A renormalization group scaling analysis for compressible two-phase flow. *Phys. Fluids A*, 5(11):2929–2937, October 1993.

[197] X. Zhou. Bridging the gap between Amdahl's law and Sandia Laboratory's result. *Commun. ACM*, 32(8):1014–1015, 1989.

[198] J. Zhu. Reservoir simulations on hypercubes. In *Proceedings of the 6th Distributed Memory Computing Conference*, pages 497–500. IEEE Computer Society Press, 1991.

[199] J. Zhu. On the implementation issues of domain decomposition algorithms for parallel computers. In R.B. Pelz, A. Ecer, and J. Häuser, editors, *Parallel Computational Fluid Dynamics '92*, pages 427–438. Elsevier Science Publishers B.V., Amsterdam, 1993.

[200] J. Zhu. A parallel time stepping algorithm on Intel iPSC/860 hypercubes. In Vichenevtsky, Knight, and Richter, editors, *Advances in Computer Methods for Partial Differential Equations*, pages 852–856. IMACS, 1993.

[201] J. Zhu. QR factorization for the regularized least squares problem on hypercubes. *Parallel Computing*, 19:939–948, 1993.

[202] J. Zhu. Solving the least squares problem on KSR-1 parallel computers. In *Proceedings of the Sixth SIAM Conference on Parallel Processing for Scientific Computing*, pages 363–366. SIAM, 1993.

[203] J. Zhu. Simulating multi-phase flow in porous media on KSR-1 parallel computers. To appear in *Applied Mathematics and Computatoins*, 1994.

[204] J. Zhu and Y.M. Chen. History matching for multiphase reservoir models on shared memory supercomputers. *Int. J. Supercomputer Applications*, 6(2):193–206, 1992.

[205] J. Zhu and Y.M. Chen. Parameter estimation for multiphase reservoir models on hypercubes. *Impact of Computing in Science and Engineering*, 4:97–123, 1992.

[206] R. Zink. The stuttgart parallel processing library SPPL and the X windows parallel debugger XPDB. In *Proceedings of the Seventh International Parallel Processing Symposium*. IEEE Computer Society Press, 1993.

Index

9 789810 215781